Biology of the Antarctic Seas XII

American Geophysical Union | ANTARCTIC RESEARCH SERIES

Volume 35

ANTARCTIC
RESEARCH
SERIES

Biology of the Antarctic Seas XII
David L. Pawson, Editor

Biogeography of Lanternfishes (Myctophidae) South of 30°S

Richard Frank McGinnis

Ⓢ American Geophysical Union
Washington, D. C.
1982

Volume 35	ANTARCTIC RESEARCH SERIES

BIOLOGY OF THE ANTARCTIC SEAS XII
DAVID L. PAWSON, *Editor*

BIOGEOGRAPHY OF LANTERNFISHES (MYCTOPHIDAE) SOUTH OF 30°S
RICHARD FRANK MCGINNIS

Library of Congress Cataloging in Publication Data

McGinnis, Richard Frank.
 Biogeography of lanternfishes (Myctophidae) south of 30°S.

 (Biology of the Antarctic seas ; 12) (Antarctic research series ; v. 35)
 Bibliography: p.
 1. Lantern-fishes—Geographical distribution. 2. Fishes—Geographical distribution. 3. Fishes—Antarctic regions—Geographical distribution. I. Title. II. Series. III. Series. Antarctic research series; v. 35.
 QH95.58.B56 vol. 12 [QL638.M9]   574.92′4s 81–14931
 ISSN 0066–4634   [597′.55]
 ISBN 0–87590–181-6

Published by
AMERICAN GEOPHYSICAL UNION
With the aid of grant DPP-8019997 from the
National Science Foundation

Printed in the United States of America

THE ANTARCTIC RESEARCH SERIES:
STATEMENT OF OBJECTIVES

The Antarctic Research Series, an outgrowth of research done in the Antarctic during the International Geophysical Year, was begun early in 1963 with a grant from the National Science Foundation to AGU. It is a book series designed to serve scientists and graduate students actively engaged in Antarctic or closely related research and others versed in the biological or physical sciences. It provides a continuing, authoritative medium for the presentation of extensive and detailed scientific research results from Antarctica, particularly the results of the United States Antarctic Research Program.

Most Antarctic research results are, and will continue to be, published in the standard disciplinary journals. However, the difficulty and expense of conducting experiments in Antarctica make it prudent to publish as fully as possible the methods, data, and results of Antarctic research projects so that the scientific community has maximum opportunity to evaluate these projects and so that full information is permanently and readily available. Thus the coverage of the subjects is expected to be more extensive than is possible in the journal literature.

The series is designed to complement Antarctic field work, much of which is in cooperative, interdisciplinary projects. The Antarctic Research Series encourages the collection of papers on specific geographic areas (such as the East Antarctic Plateau or the Weddell Sea). On the other hand, many volumes focus on particular disciplines, including marine biology, oceanology, meteorology, upper atmosphere physics, terrestrial biology, snow and ice, human adaptability, and geology.

Priorities for publication are set by the Board of Associate Editors. Preference is given to research projects funded by U.S. agencies, long manuscripts, and manscripts that are not readily publishable elsewhere in journals that reach a suitable reading audience. The series serves to emphasize the U.S. Antarctic Research Program, thus performing much the same function as the more formal expedition reports of most of the other countries with national Antarctic research programs.

The standards of scientific excellence expected for the series are maintained by the review criteria established for the AGU publications program. The Board of Associate Editors works with the individual editors of each volume to assure that the objectives of the series are met, that the best possible papers are presented, and that publication is achieved in a timely manner. Each paper is critically reviewed by two or more expert referees.

The format of the series, which breaks with the traditional hard-cover book design, provides for rapid publication as the results become available while still maintaining identification with specific topical volumes. Approved manuscripts are assigned to a volume according to the subject matter covered; the individual manuscript (or group of short manuscripts) is produced as a soft cover 'minibook' as soon as it is ready. Each minibook is numbered as part of a specific volume. When the last paper in a volume is released, the appropriate title pages, table of contents, and other prefatory matter are printed and sent to those who have standing orders to the series. The minibook series is more useful to researchers, and more satisfying to authors, than a volume that could be delayed for years waiting for all the papers to be assembled. The Board of Associate Editors can publish an entire volume at one time in hard cover when availability of all manuscripts within a short time can be guaranteed.

BOARD OF ASSOCIATE EDITORS
ANTARCTIC RESEARCH SERIES

Contents

Acknowledgments

I am most grateful to Basil Nafpaktitis, chairman of my doctoral committee, for guidance, patience, and inspiration. His perceptive and scrupulous criticism was essential to the completion of the study. Thanks are also due to Jay Savage, Ian Straughan, Giles Mead, and Barnard Pipkin, of my doctoral committee, for their criticism and suggestions. Jay Savage originally suggested the problem and guided the research in its early stages.

For providing material or facilities for the present study, I gratefully acknowledge Robert Wisner of Scripps Institution of Oceanography, Robert Lavenberg of the Los Angeles County Natural History Museum, Robert Gibbs of the U.S. National Museum, Alwyn Wheeler of the British Museum of Natural History, N. A. Mackintosh of the Whale Research Unit of the National Institute of Oceanography, and the Allan Hancock Foundation and Department of Biological Sciences of the University of Southern California.

I am indebted to the following individuals for helpful discussions of various aspects of the study: John Fitch, California Department of Fish and Game; Michael Horn, California State University at Fullerton; Ronald Heyer, U.S. National Museum; H. Geoffrey Moser, National Marine Fisheries Service, LaJolla; John Paxton, Australian National Museum; and Brent Davy, Shelly Johnson, Robert Lavenberg, William O'Day, Richard Pieper, and Theodore Pietsch, University of Southern California.

I wish to thank Jens Knudsen of Pacific Lutheran University, whose awareness, enthusiasm, and encouragement stimulated my interest in biology. My wife, Diane, provided strength and patient assistance.

For financial assistance, I am grateful to the National Science Foundation, Bernard Abbot and the Graduate School of the University of Southern California, and Pacific Lutheran University. This study was supported in part by National Science Foundation grants G-23647 and USARP GA-238, and Jay M. Savage was the principal investigator. Revision of the manuscript was supported by a Research Corporation grant to the author for studies of Southern Ocean lanternfishes. Their support is appreciated. Pacific Lutheran University has generously provided technical assistance, and I particularly wish to thank Pam Blair, Dawn Lutton, and Ann Lambert for typing the manuscript.

Abstract

A total of 84 nominal species of lanternfishes (Myctophidae) occur in the southern hemisphere south of 30°S. The distributions of these species have been analyzed in relation to the hydrology of the area. Antarctic/ Antarctic polar front, subantarctic, transitional water, and warm water lanternfish complexes, each reflecting a pattern of distribution associated with major hydrographic phenomena, have been defined and discussed. It is concluded that systems of oceanic circulation, as well as vertical distribution and inherited tolerances of these fishes to environmental variables, are important in determining their distribution. Endemic genera have been shown to occur in Antarctic-subantarctic, southern transitional, cold north Pacific, and warm water regions of the World Ocean. A consideration of paleontological and paleoceanographic literature leads to the hypothesis that conditions conducive to the evolution of these genera existed in the Oligocene era, the family achieving its present level of generic differentiation by the Miocene era. It is suggested that recent lanternfish complexes evolved from early Tertiary faunas during the marked climatic and oceanographic fluctuations of the Pliocene and Pleistocene eras.

Introduction

Perspective

With the hope of gaining ecological understanding of the midwater fauna of the Southern Ocean, an intensive sampling program was instituted in 1962 by the University of Southern California as part of the broad studies aboard the National Science Foundation sponsored vessel, USNS Eltanin. Extensive use of a 10-ft. (3-m) midwater trawl formed an integral part of this program and numerous samples of midwater organisms were acquired. Present, and frequently dominant, in most of these samples were representatives of the lanternfish family Myctophidae.

The Myctophidae, which includes about 200 species, is the most speciose family of oceanic fishes and also one of the most speciose families of oceanic organisms. Indeed, on the basis of estimates of the number of species of oceanic fishes by Parin [1970] and Cohen [1970] it would seem that myctophids comprise 15%-20% of the oceanic ichthyofauna. The family is also widely distributed, occurring in Arctic, Antarctic, and intermediate latitudes. There is evidence that myctophids are the dominant fishes in most oceanic midwaters. They comprise over 45% of the total number of fish larvae in samples examined from three oceanic areas of the Pacific and Indian oceans [Moser and Ahlstrom, 1970], and they frequently contribute the largest fraction of fish biomass in midwater trawl samples [Paxton, 1972]. There is also evidence to suggest that the family has been of similar importance in the past, at least through much of the Cenozoic era: numerous fossil myctophids are known from Quaternary and Tertiary marine deposits of North America and Europe [Fitch, 1969; Danil'chenko, 1967]. Their abundance has prompted many students, including Nafpaktitis [1968] and Kashkin [1967], to stress their importance in the trophodynamics of the oceanic ecosystem. Lanternfishes feed mainly on crustaceans [Paxton, 1967b; R. F. McGinnis, personal observation, 1974] and in turn are consumed by larger predators which are preyed upon by apex predators such as billfishes and tunas [Moser and Ahlstrom, 1970]. They may also directly provide an energy source for apex predators: Fitch and Brownell [1968] report that myctophids accounted for nearly 90% of over 18,000 otoliths found in 17 cetacean stomachs. There can be little doubt that lanternfishes represent a remarkably successful radiation of ecologically important low-level carnivores in the midwater of the World Ocean and that an understanding of their biogeography could provide significant insight into the nature and evolution of midwater communities.

Most previous studies of lanternfishes concentrate on particular taxa or particular regions of the Atlantic, Indian, Pacific, and Antarctic oceans [Bolin, 1959; Becker, 1963a, b; 1964a; 1965; Andriashev, 1962; Paxton, 1967a; Nafpaktitis, 1968; Nafpaktitis and Nafpaktitis, 1969; Moser and Ahlstrom, 1970; Backus et al., 1970; and others]. Becker [1964b, 1967a, b] presents a general survey of lanternfish biogeography, with particular emphasis on the Pacific Ocean. He discerns two major biogeographic complexes in the world ocean: a tropical warm water complex and a temperate cold water complex, the latter including northern and southern components which overlap with the tropical warm water complex at the polar edges of the subtropical anticyclonic gyres. He also summarizes patterns of distribution within these complexes. His studies, as well as others cited above, indicate that such patterns frequently conform to water masses and frontal zones as do the patterns of other pelagic organisms such as euphausiids [Brinton, 1962], epipelagic fishes [Parin, 1970], and the midwater melamphaid fishes [Ebeling, 1962]. It has become apparent that similar patterns of distribution are shown by many pelagic species, regardless of taxon, trophic level, or size [McGowan, 1971].

A rigorous analysis of lanternfish distribution has been prevented by numerous systematic problems, a lack of hydrologic and distributional data from vast geographical areas (particularly in the subtropics), and inadequate or only partially analyzed data from other large areas. However, several factors tend to alleviate such difficulties in an analysis of lanternfish distribution in the Southern Ocean. The comprehensive hydrologic and biogeographic studies which have resulted from the Discovery expeditions to the Antarctic Ocean provide a conceptual and comparative basis. Recent studies by Becker [1963a, b, 1967], Craddock and Mead [1970], Nafpaktitis and Nafpaktitis [1969], and particularly Andriashev [1962] provide valuable information on the distribution and systematics of the lanternfish fauna of the Southern Ocean and contiguous areas. Paxton [1972] and Moser and Ahlstrom [1970] have established a more rigorous definition and understanding of myctophid phylogeny. And, most importantly, the collections of the Eltanin and the Discovery expeditions provide an immense amount of data by which to conduct such a study.

Objectives of the Study

The present study proposes to describe and analyze the apparent distributions of lanternfishes in high southern latitudes of the World Ocean. The primary objectives of the study are (1) the delineation of species composition south of 30°S and their apparent distribuions in the World Ocean and (2) the definition of patterns of distribution and analysis of their relationship to hydrology and phylogeny. In addition, accomplishment of these objectives make possible a discussion of factors considered responsible for distribution and speciation of pelagic organisms, as well as of the distribution and relationships of the Myctophidae in the World Ocean.

Historical Review

More than 300 nominal lanternfish species have
been reported since the first description of a
member of this family by Rafinesque in 1810 [Pax-
ton, 1972]. Despite their importance in pelagic
ecology and an extensive interest by ichthyolo-
gists, little is known of their biology, and
only recently have relationships of and within
the family been scrutinized carefully. As is
characteristic of many midwater organisms, the
lanternfishes are capable of luminescence. Most
species have numerous photophores, as well as
accessory luminous tissue. Early systematic studies recognizing the taxonomic importance of photophore patterns culminated in the works of Bolin
[1939] and Fraser-Brunner [1949], which provided
a framework for subsequent taxonomic studies.
Fraser-Brunner [1949] suggested evolutionary
trends within the family which have since generally been confirmed by studies on osteology [Pax-
ton, 1972] and larval morphology [Moser and Ahlstrom, 1970]. Paxton recognizes 28 genera in 6
tribes and 2 subfamilies. While disagreeing with
some generic relationships proposed by Paxton,
Moser and Ahlstrom also recognize two major radi-
ations within the family. The trend in what they
term the 'narrow-eyed' subfamily has been primarily caenogenetic: the adults have remained morphologically conservative, whereas the larvae
have achieved numerous specializations,
particularly in eye morphology. A palingenetic
radiation in the second 'round-eyed' subfamily
has resulted in diversification of adult morphology, particularly of the jaws and of photophore
pattern, whereas the larvae have remained relatively unspecialized. Interestingly, an
ecological classification based on patterns of
vertical distribution and suggested by Becker
[1967b] and Parin [1970] is similar to the
proposed evolutionary models of Paxton and Moser
and Ahlstrom. Although there are many exceptions, most narrow-eyed species are members of
Becker and Parin's 'nyctoepipelagic' group, and
most round-eyed species are members of their 'deep
water' group. The similarity may be more than
casual.

Representatives of both subfamilies are known
from the Southern Ocean, and considerable literature on the myctophids of high southern latitudes
has accumulated since Richardson [1844] described
material collected by the Erebus and Terror between 1839 and 1843. This literature includes
reports on small but significant collections made
near the northern boundary of the Southern Ocean
during the Challenger Expedition [Günther, 1878,
1887], the German Deep-Sea Expedition [Brauer,
1906], and the Dana expeditions [Tåning, 1932]
and in more southern latitudes by the polar sur-
veys of the Swedish South Polar Expedition [Lönn-
berg, 1905], the German South Polar Expedition
[Pappenheim, 1912, 1914], the Scottish National
Antarctic Expedition [Regan, 1913], the Australasian Antarctic Expedition [Waite, 1916], and
British Australia New Zealand Antarctic Research
Expedition (BANZARE) [Norman, 1937]. Also included in the literature are studies on inciden-
tal and regional collections [Günther, 1864;
Gilbert, 1911; McCulloch, 1923; Barnard, 1925;
Wisner, 1963; Becker, 1963a]. Norman's [1930]
report is based on much of the Atlantic material
then available from the Discovery expeditions to
the Southern Ocean. Fraser-Brunner [1949] also
used Discovery material for his revision of the
Myctophidae. The Soviet Antarctic Expedition
provided material for the most detailed study of
myctophids to date, that of Andriashev [1962],
who emphasizes the marked differentiation, diversity, and endemism of the Antarctic fauna and
distinguishes three biogeographic regions in the
southern Pacific: Antarctic and notal regions
characterized by the myctophids Electrona antarctica and Electrona subaspera, respectively, and a
subtropical region characterized by a scombere-
socid, Scomberesox sp. Four more recent studies
add considerably to the literature on southern
myctophids: Bussing's [1965] account of the midwater fishes, including myctophids, collected by
the USNS Eltanin in the Peru-Chile Trench, Becker's [1967a] study of the myctophids collected
by the R/V Peter Lebedev Atlantic Expedition,
including material from the western Atlantic sector of the Southern Ocean, Nafpaktitis and Nafpaktitis' [1969] account of the myctophids (ex-
cluding the genera Diaphus and Lobianchia) collected by the R/V Anton Brunn during the International Indian Ocean Expedition and, finally,
Craddock and Mead's [1970] list of myctophids
collected during cruise 13 of the R/V Anton Brunn
in the southeast Pacific between central Chile
and 92°W. The data and conclusions in these
reports are considered in the present study.

Materials and Methods

The present study is based on data from the *Eltanin*, *Discovery* and *William Scoresby* collections, from miscellaneous smaller collections, and on data obtained from the literature.

Eltanin Data

Lanternfishes collected during *Eltanin* cruises 4-27 and 32-35 are included in the analysis. Larvae, unidentifiable juveniles, and damaged specimens are excluded. Most material was collected with an open 10-ft (3-m) Isaacs-Kidd midwater trawl (IKMT), a description of which may be found in Isaacs and Kidd [1952]. The remainder was collected with a 1-m IKMT, a dip net, or by bottom trawls passing through the water column. Three series of station numbers were used for labeling the collections. The University of Southern California (USC) series is the most extensive and includes samples from cruises 4-16, 18, 19, 22, 24, 26, 27, 32, 34, 35, and most samples from cruise 33. The Smithsonian Oceanographic Sorting Center (SOSC) series includes cruises 17, 20, 21, and 25. An additional series was used for cruise 33, during which some samples were labeled by ship station number, e.g., 33-1, 33-2, etc. Data for the USC series through cruise 27 (USC stations 1-1986) have been published [Savage and Caldwell 1965, 1966, 1967]. Data for the remaining stations are on file at USC, the SOSC, and the Los Angeles County Museum of Natural History. Brief descriptions of the various cruises are found in volumes 1-4 of the *Antarctic Research Journal*. The area from which samples were taken extends approximately from 30°W to 120°E.

Discovery and William Scoresby Data

A considerable portion of the myctophids collected by the *Discovery* and *William Scoresby* expeditions in the Southern Ocean is utilized in the present analysis. It includes catalogued and uncatalogued material at the British Museum of Natural History as well as unsorted material at the Whale Research Unit of the National Institute of Oceanography. Time limitations restricted the examination of many unsorted samples to a cursory glance. The bottles containing the samples were opened only if fishes were observed. Larvae and unidentifiable juveniles were not recorded. Most of the material was collected with 1-m stramin nets towed obliquely or horizontally or with 2-m nets towed obliquely. A description of *Discovery* techniques may be found in Kemp et al. [1929].

Station data are available in *Discovery* reports 1, 3, 4, 21-26, and 28.

Miscellaneous Data

Additional material from the 'Southern Ocean' in collections at the British Museum of Natural History and the Scripps Institute of Oceanography is included in the study. Data for specimens not recorded in the literature may be found in Appendix 1.

Data From the Literature

A literature survey forms an important part of the analysis. An attempt is made to consider all published records of species present in the *Eltanin* and *Discovery* collections from south of 30°S. Questionable records, including specimens inadequately described or with questionable station data, are considered in specific accounts but are not included in the analysis.

Treatment of Data

Recent systematic studies were consulted for identification of as many species as possible. However, the existing literature proved inadequate for a number of groups and a series of revisions, which have uncovered approximately 15 undescribed species, has been undertaken. Undescribed or indeterminable species are treated as '*Genus*' sp. A, sp. B, etc., and a brief description of their probable relationshps and a diagnosis are given in the respective species accounts. Material examined and included in the study is listed for each species in Appendix 1. Station number, number of specimens, and maximum and minimum standard lengths at each station are given. Published records included in the analysis are listed in the synonymies or, if from north of 30°S, referred to in the species accounts. Data for each species are summarized on a map adapted from American Geographical Society Map 3a of the *Antarctic Map Folio Series*.

Remarks on vertical distribution are based on data from USC stations 1-1986, *Discovery* stations, and the literature. The intent of such remarks is only to emphasize the least depth range at which a species is usually collected. Samples of the more common species were examined for gonadal development; the term gravid refers to females containing large ovaries filled with yolk-laden eggs. Measurements, unless otherwise stated, refer to standard length.

Description of the Study Area

The Concept of a 'Southern Ocean'

One outstanding feature in the high latitudes of the southern hemisphere is a circumpolar continuity of the prevalent atmospheric and associated hydrospheric phenomena. Strong westerly winds south of the subtropical anticyclonic gyres of the Atlantic, Indian, and Pacific oceans result in the continuous movement of water that surrounds the Antarctic continent. This vast current, the west wind drift, is bound to the south by the east wind drift, a smaller and fragmented current adjacent to the Antarctic coast that is driven to the west by the polar easterlies. The northern limits of the west wind drift are difficult to define, measure, or observe. The region where waters of the west wind drift confront subtropical waters seems to be so variable and in some places, particularly the southeastern Pacific, so obscure that some authors choose to ignore the concept of a 'Southern Ocean' and include the waters of high southern latitudes in their definitions of the Atlantic, Indian, and Pacific oceans [Viglieri, 1966; Gordon, 1971a; international agreement, Ostapoff, 1965, p. 97; and others]. The U.S. Board on Geographic Names has studied the matter on a number of occasions. It does not recognize a 'Southern Ocean' and has consistently recommended that the names Atlantic Ocean, Pacific Ocean, and Indian Ocean be applied southward to the coast of Antarctica. This recommendation by the Board has been followed in other volumes of the Antarctic Research Series. Others recognize a 'Southern Ocean' but use a variety of definitions for its northern boundary: the approximate latitude of the Antarctic polar front (Antarctic convergence), where westerly winds are maximal [Herdman, 1966], regions 'north of the Antarctic polar front' [Hasle, 1969], the subtropical convergence as drawn by Deacon [1937] [see Roper, 1969; Knox, 1970], and the northern limit of the west wind drift as arbitrarily defined by a line approximating the subtropical convergence and drawn from definite points on land masses [Kort et al., 1965]. Inadequate data and incomplete understanding of the dynamics of the region of the subtropical convergence preclude any rigid, sharp definition of the northern limit of the 'Southern Ocean'. The northern limit of influence of the west wind drift does not coincide with the southern limit of influence of the anticyclonic gyres [Boltovskoy, 1968]; that is, a region of transition exists between the two circulations. I accept the concept of a 'Southern Ocean' and consider its northern boundary to be the region of transition near the subtropical convergence. Conversely, I consider the same region as the southern limit of the Atlantic, Indian, and Pacific oceans.

The Model Southern Ocean

The hydrology of the 'Southern Ocean' as interpreted by Deacon [1933, 1937] is summarized, sup-

ported, defined, and partially challenged or ignored in more recent literature. Sverdrup et al. [1942], Mackintosh [1946], Midtun and Natvig [1957], Wyrtki [1960a, b], Kort [1962], Deacon [1963], Reid [1965], Botnikov [1966], Gordon [1967, 1971a, b], Gordon and Goldberg [1970], Warren [1970], and many others deal partly or entirely with the 'Southern Ocean' as described by Deacon. _Eltanin_ hydrologic data for Antarctic waters between 20°W and 170°W are analyzed by Gordon [1967, 1971a, b]. Gordon and Goldberg [1970] summarize available hydrologic data for the oceans which surround the Antarctic continent in a series of horizontal and vertical sections of temperature, salinity, and oxygen distributions. A brief description of the model hydrology of the 'Southern Ocean' from these sources and a consideration of modifications of, or exceptions to, the model follows. The terminology is one widely used by oceanographers.

Meridional and vertical movements of water are imposed on the zonal flow of the east and west wind drifts. Meridional components comprise the balance of flow into and out of the 'Southern Ocean' and integrate, to some degree, the waters of the Atlantic, Indian, and Pacific oceans. Figure 1 qualitatively illustrates the movements of water. The major flow of water into the 'Southern Ocean' involves deep water flowing from the north (primarily from the North Atlantic [Gordon, 1971b]). The deep water entrains adjacent waters, upwells as Antarctic circumpolar water (ACW) at the Antarctic divergence, and mixes with and forms Antarctic surface water (ASW). The divergence marks the zone between northeastward-flowing ASW of the west wind drift and southwestward-flowing ASW of the east wind drift. The latter contributes to the formation of Antarcic bottom water (ABW), which flows beneath and entrains ACW from source areas near the continent. The northern limit of ASW is marked by one of the most striking features of the World Ocean, a zone of surface temperature gradient which circumscribes the earth near 52°S. This zone, the Antarctic convergence, or, more correctly the Antarctic Polar Front (APF), is the area of confrontation of the northward-flowing ASW and the less dense subantarctic surface water (SASW) into which some ASW is incorporated. The remainder of the ASW sinks and flows northward over the deep water as Antarctic intermediate water (AIW). There may be a considerable return flow of AIW into the ACW [Pytcowitz, 1968]. The SASW flows northward and mixes with subtropical surface water (STSW) of the anticyclonic gyres at the subtropical convergence, or the region of transition [Deacon, 1937; Gordon, 1971a]. The dynamics of this region are not understood, and disagreement exists as to whether SASW [Deacon, 1937; Heath, 1968; and others], STSW [Wyrtki, 1960b; Rotschi and Lemasson, 1967; and others], or a mixture of the SASW and STSW [Schell, 1968] and a southward-flowing current between AIW and

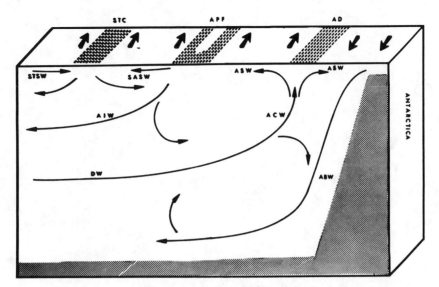

Fig. 1. Schematic representation of the hydrology of the Southern Ocean. Arrows indicate direction of flow. ABW is Antarctic bottom water; ACW, Antarctic circumpolar water; AIW, Antarctic intermediate water; ASW, sub-antarctic surface water; STSW, subtropical surface water; AD, Antarctic divergence; APF, Antarctic Polar Front; and STC, subtropical convergence. See text.

SASW [Deacon, 1937; Heath, 1968] are formed in this region.

Gordon [1967, 1971a] characterizes Antarctic water masses by extrema layers of a variety of physical parameters. A subsurface temperature minimum of 0.8°-1.6°C varying in depth from 50 to 300 characterizes the ASW throughout most of the year. Winter cooling of the surface layer obliterates the subsurface minimum. The ASW is shallowest at the Antarctic divergence, gradually deepening towards the north until it reaches the vicinity of the APF, where the slope of the minimum increases drastically and disappears. The ACW is composed of two layers: an upper layer characterized by a temperature maximum of 0.2°C-3°C and an oxygen minimum of 3.9-5.0 ml/l and a lower layer characterized by a salinity maximum of 34.67-34.75 per mil. Both layers slope upward to the south and are shallowest at the Antarctic divergence. The ABW is characterized by a deep oxygen maximum and a potential temperature minimum, both of which originate from a mixture of cold shelf water and deep water. Mixed waters occur between all core layers.

A number of surface and subsurface parameters are used to define the position of the Antarctic Polar Front. Surface temperature gradient is most commonly used and most easily detected, but subsurface parameters are preferred [Gordon, 1971b]. The presence of a double frontal structure is common. A southern secondary frontal zone characterized by a northward increase in surface temperature and depth of the temperature minimum layer is separated from a northern primary frontal zone characterized by a more marked increase in surface temperatue and a vertically elongate temperature minimum layer by a warm divergent zone of relatively stable water in which the temperature minimum layer is weak or absent [Gordon, 1967, 1971b]. The SASW is a vertically extensive layer of water north of the APF. It is characterized by a temperature of more than 3°C and a uniform salinity structure of about 34.0-34.5 per mil. A subsurface salinity maximum in more northern subantactic latitudes suggests a southward flow of subsurface water. A mixed layer lies between the SASW and the AIW. Well to the north of the APF, the AIW is characterized by a salinity minimum at 900-1000 m. A relatively marked longitudinal temperature gradient of about 4°C, the subtropical convergence, is frequently used to delimit the northern limit of SASW and temperatures of more than 12°C and salinities of about 35.0 per mil are indicators of surface water masses north of the STC [Deacon, 1937].

Figure 2 shows the major surface hydrographic features of the study area, including the regions of the Polar Front and subtropical convergence as well as the Weddell-Scotia confluence and regional currents (discussed below). In the Pacific sector, the position of the Polar Front is based on Gordon's [1971a] intepretation of Eltanin data. In the remaining sectors a band, 3° of latitude wide, has been imposed in the position of the Polar Front drawn by Mackintosh [1946]. The position of the subtropical convergence is placed within a sharp north to south salinity change which separates uniformly dilute subantarctic waters from uniformly saline subtropical waters. It includes the line of the subtropical convergence drawn by Deacon [1937] from relatively few surface temperature measurements. Subsurface isohalines were chosen to dampen seasonal effects and increase the comparability of data taken from a variety of sources [Wüst, 1936; Deacon, 1937; Jacobs, 1966; Jacobs and Amos, 1967; Scripps Institution of Oceanography, Woods Hole Oceanographic Institution, 1969; Craddock and Mead, 1970]. The precise northern and southern limits of the region of the subtropical convergence were arbitrarily placed at the positions of the 34.8 and 34.6 per mil isohalines, respectively. These

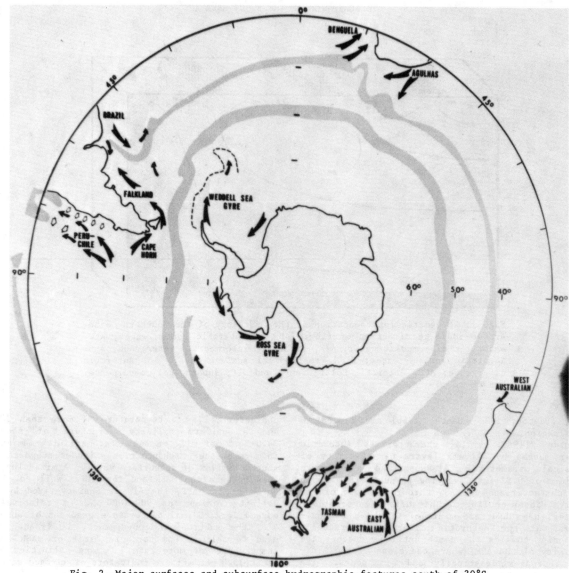

Fig. 2 Major surfaces and subsurface hydrographic features south of 30°S.
The regions of the Polar Front and subtropical convergence are delimited
by the southern and northern shaded areas, respectively. The Weddell-
Scotia confluence is delimited by the dashed line. Regional currents are
labeled and schematically represented by arrows. See text for additional
discussion.

isohalines probably more closely approximate the southern limits of subtropical waters rather than the entire region of transition between subantarctic and subtropical waters. The position of the Weddell-Scotia confluence is taken from Gordon [1967].

Pacific Sector

The presence of the South American continent in the course of the west wind drift results in a divergence of subantarctic waters near the Chilean coast. This water forms into northward- and southward-flowing currents in the vicinity of the weak and variable winds of the horse latitudes. This divergence has been variously placed near 50°S [Sverdrup et al., 1942; Defant, 1961; Wooster and Reid, 1963; Wooster, 1970], near 45°S [Deacon, 1937; Dietrich, 1963], and near 40°S [Muromtsev, 1963]. Guenther [1936] places the origin of a

northward coastal current near 40°-41°S, except in winter when it supposedly originates near 33°S. The disagreement is probably attributable to one or a combination of the following factors: paucity of data, local processes such as upwelling that obscure large-scale dynamics when data are scarce and, possibly, the existence of a broad and variable region of divergence. Regardless of the disagreement, currents which originate between 40°S and 50°S evidently flow north of 40°S and south of 50°S near the Chilean coast. Isotherms and isohalines which are zonal in surface and sub-surface layers west of Chile and between 40°S and 50°S near the coast parallel the coast in areas influenced by these currents. The relatively strong and narrow southward flow moves around South America into the Atlantic as the Cape Horn Current [Boisvert, 1967].

The Peru Current system to the north of the

divergence is much more complex. A barely distinct boundary separates a northward-flowing narrow coastal current from a broader oceanic current [Guenther, 1936; Wyrtki, 1963]. A divergence of these currents near 25°S effects a southward flow of oxygen-depleted and saline waters in a countercurrent [Wyrtki, 1963, 1967]. A second southward current flows beneath the coastal current as compensation for coastal upwelling [Guenther, 1936] to at least 41°S [Wooster and Gilmartin, 1961]. This current appears in Eltanin hydrologic data as a subsurface salinity maximum extending as far south as 46°S near the coast [Friedman, 1964]. Wooster [1970] suggests that the multicurrent model for the Peru Current system indicates a lack of understanding. He also suggests that a current originating south of 40°S separates from the coast before reaching 20°S and that the Peru Current originates along the southern coast of Peru, being fed by waters from deeper layers and waters from the west. A warm and relatively stable zone apparently separates the two currents. Craddock and Mead [1970] report the concentration of northward movement in two bands near 34°S, a larger but narrow component 350 km offshore and a wider but smaller component 750 km offshore. They found southward flow concentrated in a narrow coastal flow with less transport between the two northward currents and west of 88°W.

Antarctic intermediate water lies betwen 600 and 900 m near the coast as is evidenced by a salinity minimum (Eltanin data [Friedman, 1964]). Its presence here may be due to horizontal mixing [Wyrtki, 1963]. According to Reid [1965], vertical diffusion rather than, or in connection with, the actual sinking of water may account for the salinity minimum of AIW which flows equatorward from the west wind drift in the eastern Pacific Ocean. Taft [1963] attributes the penetration of undiluted AIW with the subtropical anticyclonic circulation to the obstruction of circumpolar flow by the American land mass. The westward extension of the subtropical gyre carries modified AIW and tropical surface and subsurface water from the east and north into the Tasman Sea [Rochford, 1960; Wyrtki, 1960b; Taft, 1963]. Very little AIW enters the Tasman Sea from the south [Taft, 1963]. The western boundary of the subtropical gyre, the East Australian Current extends to depths of at least 1100 m and possibly 2000 m [Newell, 1966]. This current might actually represent the shoreward flow of a series of anticyclonic eddies carrying water from the Coral Sea southward [Hamon, 1970]. Such a scheme might explain discrepancies in estimates of the latitude at which the East Australian Current turns eastward (34°S [Hamon, 1965], 40°S [Rothschi and Lemasson, 1967], and the southeast corner of Australia [Newell, 1966]) and account for disagreement about the position of the subtropical convergence. Deacon [1937] and New Zealand workers [e.g. Burling, 1961] extend the convergence from the vicinity of Tasmania to southern New Zealand, whereas others [Wyrtki, 1960a; Rothschi and Lemasson, 1967] extend it to northern New Zealand. The confrontation of two branches of the East Australian Current (from separate anticyclonic movements) might explain the northern zone of convergence observed by Wyrtki and others [Garner, 1967a]. Southern New Zealand appears to

be surrounded by mixed subtropical and subantarctic waters. A branch of the trade wind drift, the East Cape Current, flows south along the east coast of New Zealand and forms an anticyclonic eddy north of the convergence east of Cape Palliser [Garner, 1967b]. The influence of the South Pacific subtropical gyre extends southeast of New Zealand and eastward across the Pacific between 40°S and 50°S at depths exceeding 500 m [Gordon and Goldberg, 1970]. Tongue-shaped isopleths of relatively high temperature salinity and oxygen at 20, 200, and 500 m extend for considerable distance southeastward from the region of the convergence [Gordon and Goldberg, 1970, charts 2, 3, and 4]. The distribution of isohalines in the upper 500 m of Pacific Subantarctic waters is interpreted by McGinnis [1974] as evidence for an endemic counterclockwise circulation, the westward component of which occurs at 40°S-45°S.

Two types of ASW are separated by the secondary Polar Front in the Pacific sector [Gordon, 1971a]. According to Gordon, this front apparently represents the northern boundary of a poorly understood cyclonic gyre extending from the Ross Sea, the eastern and western limits of which are not known. The extent to which the east wind drift contributes to this cyclonic movement of water is not mentioned in the literature. Gordon [1971b] discusses possible causes of the spatially variable yet patterned structure of the Antarctic polar front in the Pacific sector but does not consider how the structure might affect the relative or absolute amounts of AIW or SASW produced in the frontal zone. Are there longitudinal variations in the amounts of AIW or SASW produced at the front that relate to its structure? Although an answer is not readily available, inspection of the charts in Gordon and Goldberg's [1970] atlas indicates that there are longitudinal variations in physical characteristics of water immediately north of the polar front and that the variability might be related to the structure of the frontal zone.

Atlantic Sector

There is a strong flow of subantarctic water into the Atlantic sector through the Drake Passage. This current includes waters of the Cape Horn Current as well as unmodified water from the west wind drift. The current, after passing through the Drake Passage, turns and bifurcates [Deacon, 1937]. The western branch, the Falkland, or Malvinas, Current, appears to receive additional water from elsewhere and proceeds northward as a deep-reaching current pressing against the continental margin of southern Argentina [Gordon and Goldberg, 1970]. The eastern branch flows northeastward and then eastward [Deacon, 1937]. Prior to turning east it meets and mixes with the irregular terminal region of the Brazil Current, which carries warm and saline waters from equatorial and central areas. Unmixed subtropical water does not seem to reach beyond 30°S-36°S, but mixed water extends as far as 47°S-49°S [Boltovskoy, 1968]. East of the Brazil Current the west wind drift and the southern part of the subtropical gyre maintain their identities and flow eastward aross the Atlantic. There is a well-defined northern boundary of SASW at about 37°S [Deacon, 1937]. The eastern limb of the subtropical gyre, including the coastal Benguela

Current, carries insignificant amounts of surface or subsurface subantarctic water [Deacon 1937]. The subtropical anticyclonic gyre is deeper than was previously thought, however, and extends through the AIW according to Taft [1963]. Despite the absence of subantarctic waters in the South Atlantic gyre, a comparison of temperature profiles in Muromtsev [1963] and Wüst [1936] does not reveal any major differences in temperatures between the southeastern Pacific Ocean and the Atlantic Ocean at depths of 200-1000 m.

South of the Polar Front, water passes through the Drake Passage into the Atlantic sector and meets with colder east wind drift water flowing north and east from the Weddell Sea, forming the Weddell-Scotia confluence (Figure 2). The confluence, which is observable at all depths of the circumpolar water and is tubulent in the surface layers [Gordon, 1967], is the northern boundary of a clockwise gyre which extends across the Atlantic sector, possibly as far as 20°-40°E, where it may turn southward [Deacon, 1937]. It bends sharply northward near the Scotia Arc at 30°W and the northeast coast of South Georgia is washed by water moving in a northwestward direction and carrying a significant Weddell Sea component [Deacon, 1937]. A northwestward bend of the Polar Front at 40°-50°W suggests a continuation of north-northwestward movement of water beyond South Georgia, perhaps even beyond the APF, where it may contribute to the unknown component of the Falkland Current postulated by Gordon and Goldberg [1970]. Deacon [1937] found evidence of a strong northward movement of Antarctic water beneath the southeastward-flowing SASW between the Falkland Islands and South Georgia. The strong northward movement of Antarctic waters in this area is evidenced in the 200- and 500-m isopleths drawn by Gordon and Goldberg [1970, Plates 4 and 5]. The southeastward surface current apparently continues as far as the southwestern coast of South Georgia, which is bathed by water flowing from the Drake Passage [Deacon, 1937].

Indian-Australian Sector

A region of transition separates the Atlantic and Indian oceans. The Agulhas Current, carrying Indian Ocean tropical and subtropical waters as an extension of the tropical convergence [Orren, 1966], flows south along the east coast of Africa and splits into a westward and a southward branch [Deacon, 1937]. The westward branch mixes with the South Atlantic subtropical gyre near South Africa and contributes water to the Atlantic through much of the year [Schell, 1968; Shannon, 1966; and others]. The southward branch turns east near 40°S, where it mixes with the west wind drift, pushing the subtropical convergence to the south near 20°E. Irregular currents and sharp gradients of isopleths characterize this area [Deacon, 1937]. The remainder of the Indian-Australian sector is characterized by Deacon as follows. Steep gradients of temperature and salinity separate the west wind drift from the Indian subtropical circulation in the central region. Further east, waters from both systems mix and spread south of Australia, the subtropical circulation apparently diverging into northward and southward components near Australia. The

southward subantarctic subsurface current flows with more strength and at greater depths than in other sectors of the 'Southern Ocean'. Waters south of the Polar Front are east of the influence of the Weddell Sea gyre and the 1° and 2°C isotherms are found further south of the APF than further west. The mixed and saline water which spreads eastward from the eastern Indian Ocean south of the subtropical convergence is termed Australasian subantarctic water by Burling [1961], the term circumpolar subantarctic water being retained for the remaining areas between the subtropical convergence and the Antarctic Polar Front.

Distributional Aspects of Plankton in the Antarctic Ocean

Despite striking temporal and spatial variability of phytoplankton composition, standing crop, and productivity in the 'Southern Ocean,' generalizations can be drawn. Antarctic waters are generally more productive and harbor greater quantities of phytoplankton than subantarctic waters. The region of the Antarctic Polar Front is characterized by low productivity and standing crop [Hasle, 1969; El-Sayed, 1970]. Maximal densities of phytoplankton occur seasonally in discrete latitudinal bands which move southward during the summer months [Hart, 1942; Hasle, 1969]. Very high productivity is characteristic only of neritic waters [El-Sayed, 1970]. Annual production in Antarctic waters which is concentrated into half of the year, with periods of increase in spring and autumn [Foxton, 1956], is comparable to that of temperate waters [Walsh, 1969]. Oceanic areas north of the subtropical convergence, however, appear to have lower rates of production than areas farther south [Koblentz-Miske et al., 1970, Figure 1].

Evidently, most Antarctic phytoplankters are circumpolar [Baker, 19954], diatoms being the dominant forms [Hasle, 1969; Balech, 1970]. Hasle's [1969] study on the phytoplankton of the 'Southern Ocean' is limited mostly to Antarctic waters of the Pacific sector. She concludes that the flora is composed of heterogeneous cosmopolitan and endemic groups, the latter being more numerous in the Pacific sector than in other oceans. This high degree of endemism in some taxa is substantiated by Balech [1970]. Despite unresolved problems, Hasle also determined that (1) the region of the subtropical convergence apparently approximated the southern distribution boundary of a complex of about 20 warm water phytoplankton species which were collected only in her northernmost subantarctic stations, (2) the region of the APF marks the southern boundary of a complex of about seven cosmopolitan and one endemic subantarctic species, (3) five diatom species endemic to the 'Southern Ocean' show a southern limit between the APF and the ice border, (4) about 23 diatom species, also endemic to the 'Southern Ocean', have their northern limit of distribution in the Antarctic waters, the majority (about 20) of which are restricted to southernmost Antarctic waters, and (5) about half of more than 20 taxa found in all parts of the study area are cosmopolitan species, the remainder apparently being endemic 'Southern Ocean' species. El-Sayed

[1970] suggests that a number of phytoplankton species may be endemic to the region of the APF.

According to Foxton [1956], the waters of the 'Southern Ocean' are characterized by higher standing crops of zooplankton than tropical or subtropical waters. In Antarctic waters, the bulk of the biomass is concentrated in the upper 100 m during the summer and between 500 and 1000 m during the winter. However, the total biomass in the upper 1000 m does not vary significantly with seasons [Foxton, 1956]. The vertical displacement is due to seasonal vertical migration of many common zooplankters. Two types of zonal maxima of mesoplanktonic biomass occur within the 'Southern Ocean' according to Voronina [1968]. The first type occurs in northern regions during the spring and moves southward following the zone of phytoplankton abundance. It correlates with the life cycle of dominant herbivorous copepods. The second type results from a mechanical concentration of plankters by water movements at the Antarctic Polar Front and, to a lesser extent, the Antarctic divergence. Herbivore densities are greatest in areas with intermediate phytoplankton densities and lowest at maximum concentrations of phytoplankters [Hardy and Guenther, 1935; Knox, 1970]. The distribution of mesoplankton biomass is similar in the Atlantic, Indian, and Pacific sectors of the 'Southern Ocean' [Voronina and Maumov, 1968]. Euphausia superba may constitute as much as one half of the zooplankton biomass in the upper 50 m of the Antarctic region. In the upper 500 m, copepods constitute 60% and 70% of the mesoplanktonic biomass in subantarctic and Antarctic waters, respectively, and only 50% in tropical waters [Voronina and Naumov, 1968].

The literature on the distribution of zooplankton in the 'Southern Ocean' is extensive. It includes studies on chaetognath [David, 1955, 1958, 1963], copepods [Andrews, 1966; Voronina, 1968], euphausiids [John, 1936; Baker, 1959, 1965; Marr, 1962; Brinton, 1962], amphipods [Kane, 1966; Hurley, 1969], salps [Foxton, 1961, 1966], polychaetes [Tebble, 1960], foraminiferans [Be', 1969], and the cephalopod genus Bathyteuthis [Roper, 1969]. These studies vary markedly in perspective and techniques, thus making comparison of data and results very difficult. Many of them examine distributions in particular sectors of the 'Southern Ocean' or along particular meridians, while most circumpolar studies are limited to single species or are based on relatively few data. No recent studies consider speciose taxa throughout the 'Southern Ocean.' Nevertheless, certain features of zooplankton distribution seem to be well established.

Again, one of the most striking features is the marked endemism among 'Southern Ocean' zooplankton. Relatively few species occur on both sides of the subtropical convergence. Species which do occur on both sides usually extend only a short distance or decrease markedly in abundance on one or the other side of the convergence. This is apparent from data summarized by Mauchline and Fisher [1969, pp. 336, 341, 344] on euphausiids, by Frost and Fleminger [1968, pp. 83-86] on clausocalanid copepods, by Foxton [1961, p. 28] on salps, and by Be' [1969, p. 10] on foraminiferans. Tebble [1960] reports that a number of pelagic polychaetes have a southern limit but that none have a northern limit at the subtropical convergence in the Atlantic Ocean. However, two species, Travesioopsis coniceps and T. levinsini, do not appear from his figures to extend more than a short distance north of the 'line' he accepts as the subtropical convergence. A high proportion of amphipod species are apparently unaffected by the convergence [Hurley, 1968], yet a considerable number of them do show a northern or a southern limit in the transition region [Hurley, 1969; Kane, 1966; Shih, 1969]. A number of zooplankters that straddle the convergence are restricted to a narrow latitudinal belt in the southern hemisphere and are termed transitional species by Brinton [1962] and McGowan [1971].

Most Antarctic, and possibly subantarctic, zooplankton species are circumpolar. Salpa gerlachei Foxton, which is limited to Antarctic waters of the Pacific sector and the Peru-Chile Trench, is an exception [Foxton, 1966]. Hydrologic boundaries which correspond to faunal boundaries for many of the circumpolar species are the edge of the ice shelf, the Antarctic divergence, and, particularly, the Antarctic Polar Front. The Polar Front is frequently used to divide the 'Southern Ocean' into Antarctic and subantarctic biogeographic provinces.

Species Account

The Myctophidae and the closely related midwater family Neoscopselidae are apparently derived from an ancestral stock which also gave rise to the sublittoral-bathyal family Chlorophthalmidae [Paxton, 1972; Moser and Ahlstrom, 1970]. Paxton [1972] and Moser and Ahlstrom [1970, 1974] recognize two subfamilies of Myctophidae, Myctophinae and Lampanyctinae. The 31 genera recognized by Moser and Ahlstrom [1974] are listed below.

Subfamily Myctophinae	Subfamily Lampanyctinae
Protomyctophum	Notolychnus
Electrona	Taaningichthys
Metelectrona	Lampadena
Benthosema	Dorsadena
Diogenichthys	Lepidophanes
Hygophym	Bolinchthys
Myctophum	Ceratoscopelus
Symbolophorus	Lampanyctus
Loweina	Parvilux
Tarletonbeania	Stenobrachius
Gonichthys	Triphoturus
Centrobranchus	Lobianchia
	Diaphus
	Lampanyctodes
	Gymnoscopelus
	Notoscopelus
	Lampichthys
	Scopelopsis
	Hintonia

Six of these genera are not represented in the study material. They include Stenobrachius (two species), Tarletonbeania (one species), and Dorsadena (one species), which are limited to the North Pacific Ocean [Becker, 1964a; 1967b; Coleman and Nafpaktitis, 1972], Centrobranchus (four species), which is distributed in warm waters of the Atlantic, Indian, and Pacific oceans [Becker, 1964a; 1967b], Triphoturus (three or four species), which seems to be associated with equatorial waters, and Parvilux (two species), one species apparently restricted to the eastern equatorial Pacific and the other apparently restricted to mixed eastern equatorial Pacific and California current waters [Hubbs and Wisner, 1964; Paxton, 1967a]. One species each of Triphoturus and Centrobranchus, however, has been reported from south of 30°S. Centrobranchus nigroocellatus (Günther, 1873), reported from just south of 30°S near 90°W [Craddock and Mean, 1970], is known also from between 35°N and 30°S in the Atlantic Ocean [Bolin, 1959], between 5°S and 30°S in the Indian Ocean [Becker, 1964a; Nafpaktitis and Nafpaktitis, 1969], and from near 30°S in the Tasman Sea [Becker, 1964a]. Triphoturus mexicanus (Gilbert, 1891), reported from south of 30°S near Chile [Bussing, 1965], may actually be a different related species [Moser and Ahlstrom, 1970; p. 114] of a complex restricted to equatorial waters of the eastern Pacific Ocean. The remaining lan-

ternfish genera are represented by 77 species in the material available for the present study and by 5 additional species which have been reported in the literature from south of 30°S. Station records for the former species are listed in Appendix 1.

Protomyctophum Fraser-Brunner, 1949

The genus Protomyctophum includes 16 species in two subgenera. According to Paxton [1972] and Moser and Ahlstrom [1970] it is closely related to Electrona.

All five described and three undescribed species of P. (Protomyctophum) are present in the study material. This subgenus is known only from the 'Southern Ocean.'

Protomyctophum (P.) anderssoni (Lönnberg, 1905)
Fig. 3

Scopelus antarcticus Boulenger, 1902, p. 174 (Victoria Land, Southern Cross).

Myctophum anderssoni Lönnberg, 1905, p. 61, (east of Falkland Islands).--Norman, 1930, p. 320 (Discovery (D) 64).--Whitley, 1941, p. 124 (Macquarie Island).

Myctophum tenisoni Norman, 1930, p. 321, partim, (D 114, 239).

Electrona (P.) anderssoni Krefft, 1958, p. 253 (northwest of Bouvetøya).

Electrona anderssoni Andriashev, 1958, p. 201, (northeast of Macquarie Island).--Barsukov and Permitin, 1959, p. 382 (northwest of Prince Edward Island).

Protomyctophum (P.) anderssoni Andriashev, 1962, p. 226 (OB 70, 269, 362, 363, 413, 442, 464; Slava 58, 70, 74, 77, 78, 90).--Becker, 1963a, p. 16; 1967a, p. 89 (near 41°S, 39°W).

Size. Protomyctophum anderssoni attains a length of about 70 mm. Gavid females as small as about 50 mm long were found in samples collected from October to January.

Distribution. Figure 3 illustrates the known distribution of P. anderssoni. It has been collected in a circumpolar band centered in the Antarctic Polar Front. This band is approximately 20° of latitude wide in the area sampled by the Eltanin (30°W to 120°E). Here, the southern limits of distribution closely correspond to the Weddell-Scotia confluence and to the Antarctic divergence as drawn by Deacon [1937]. Protomyctophum anderssoni does not occur in the water of the east wind drift that flows out of the Ross and Weddell seas. It has been collected closest to the continent in the vicinity of Victoria Land and Wilkes Land where the east wind drift is narrowest. The relatively narrow band of records near the Polar Front from 30°W east to 120°E possibly reflect the pattern of sampling. However,

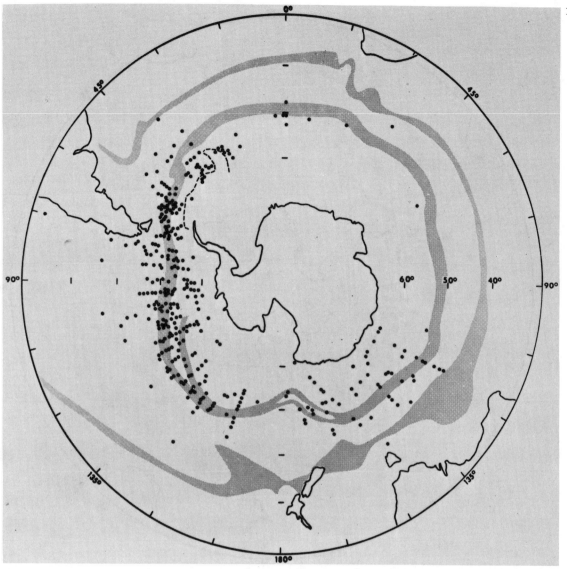

Fig. 3. Distribution of _Protomyctophum anderssoni._

the fact that other species have been collected
farther south in this same area indicates that P.
anderssoni may be rare or even absent in the eastward extension of the Weddell gyre. The northern
limits of distribution generally occur south of
the subtropical convergence. The northernmost
records are of juveniles. The juvenile reported
by Andriashev [1962] from near 34°S off Chile
represents the only record for the species from
that area, in spite of numerous trawls taken in
that vicinity.

Andriashev [1962] suggests that P. anderssoni
does not occur at depths exceeding 500 m. However, Becker [1963a] reports specimens from waters
deeper than 600 m. Eltanin material includes
juveniles and adults collected in daytime and
nighttime trawls that did not exceed 100 m. A
single specimen was collected at the surface by
the William Scoresby (British Museum 1948-5-
4-673).

Systematic notes. Protomyctophum anderssoni is
very distinct morphologically from other members
of the genus [Andriashev, 1962; Paxton, 1972] and
is presumably not closely related to any recent
species. H. G. Moser (personal communication,

1974) believes that its unique larval as well as
adult morphology may warrant generic distinction.

Protomyctophum (P.) tenisoni (Norman, 1930)
Fig. 4

Myctophum tenisoni Norman, 1930, p. 321 (part, D
65, 109, 207).
Protomyctophum (P.) tenisoni Andriashev, 1962, p.
229 (OB 455, 461).

Size. Protomyctophum tenisoni attains a length
of about 50 mm. Gavid females as small as 42 mm
long were found in samples collected from November
to January.
Distribution. The known distribution of P.
tenisoni is shown in Figure 4. Fraser-Brunner
[1949] has indicated that five of the 25 juveniles
reported by Norman [1937] from near 43°S, 94°E
were actually P. bolini. I was unable to find
the remaining specimens at the British Museum,
and because this species may easily be confused
with other forms, Norman's record is not included
in Figure 4. Protomyctophum tenisoni has a circumpolar distribution in the waters of, and ad-

12

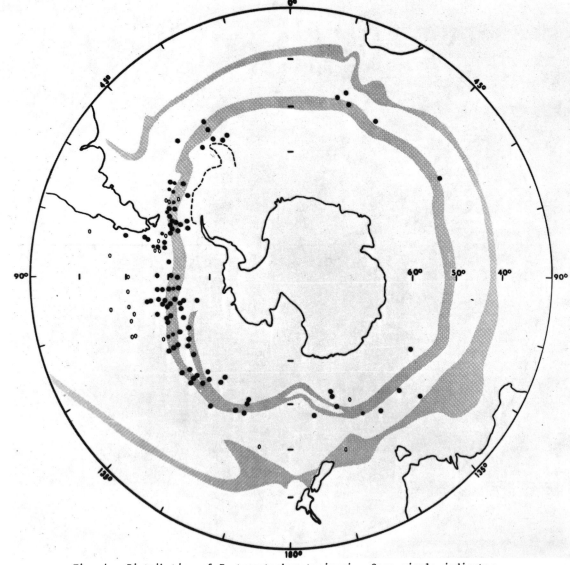

Fig. 4. Distribution of *Protomyctophum tenisoni*. Open circle indicates individuals less that 20 mm long, large dot indicates individuals more than 19 mm long, and small dot indicates individuals of both size ranges in the same sample.

jacent to, the region of the Polar Front except near South Georgia. Juveniles smaller than about 20 mm long extend farther north than, and not as far south as, larger individuals. The frequency of capture in the Australian sector and between 90°W and 105°W indicates a more than casual occurrence of this species north of the region of the Polar Front in those longitudes. The absence of records in much of the Atlantic and Indian sectors may be due to inadequate sampling.

Eltanin and Discovery material includes specimens of P. tenisoni that were collected at depths not exceeding 100 m. There is some indication that juveniles may occur deeper than adults.

Systematic notes. Protomyctophum tenisoni can be distinguished from the very closely related P. sp. A by the presence of pigment on its caudal peduncle, a more posterior position of its SAO_1 photophore, and, usually, the lower number of gill rakers on the first gill arch (25 versus 26-27), and its higher number of AO photophores

(17-18 versus 16-17). Adult males can also be distinguished from sympatric adult males of P. sp. A by their smaller size and the morphology of the supracaudal luminous gland.

Protomyctophum (P.) sp. A
Fig. 5

Myctophum tenisoni Norman, 1930, p. 321 (part, D 44).

Size. Protomyctophum sp. A, which is composed of two populations of uncertain status (see the systematic notes), attains a length of about 80 mm.

Distribution. The known distribution of Protomyctophum sp. A is shown in Figure 5. The specimens illustrated as P. normani by Fraser-Brunner [1949] from the 'Southern Ocean' and by Angelescu and Cousseau [1969] from the stomach of a hake collected on the Argentine shelf are P. sp. A.

13

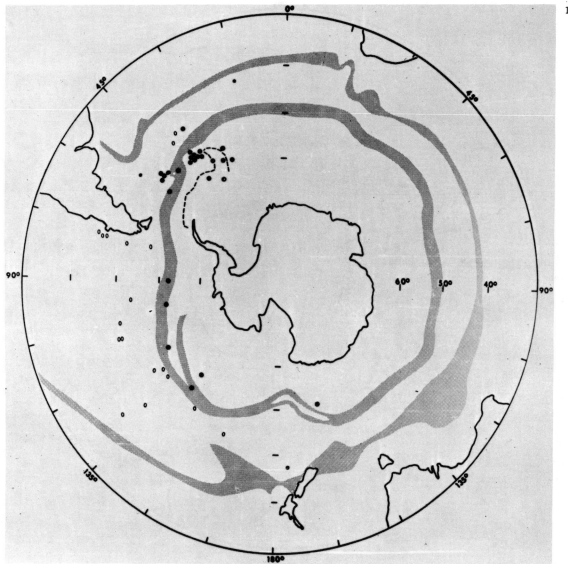

Fig. 5. Distribution of <u>Protomyctophum</u> sp. A. Open circle indicates
individuals less than 30 mm long, small dot indicates individuals 37-42 mm
long, and large dot indicates individuals larger than 58 mm long.

These are not included in Figure 5 because specific localities are not given. Juveniles less
than 30 mm long were collected north of the Polar
Front. Individuals larger than 55 mm long were
taken in or near the region of the Polar Front in
the Pacific sector and north and south of it in
the southwestern Atlantic sector. The absence of
records from most of the eastern hemisphere may
reflect inadequate sampling, although numerous
negative <u>Eltanin</u> stations south of Australia indicate its absence there. The lack of records
from the Drake Passage, where <u>P. tenisoni</u> seems
to be common, indicates that the population in-
habiting the area between the Argentine shelf and
south of the Weddell-Scotia confluence maintains
itself wihout recruitment from the Pacific population. It may be maintained by clockwise currents extending around South Georgia.

Young <u>P.</u> sp. A evidently occur deeper than
adults do. Specimens less than 26 mm were collected only in trawls which fished deeper than 200

mm, whereas larger specimens were repeatedly taken
by the <u>Discovery</u> at depths less than 200 m. One
adult male was collected at the surface (Elt
(Eltanin) 1524).

<u>Systematic notes</u>. <u>Protomyctophum</u> sp. A can be
distinguished from the very closely related <u>P.</u>
<u>tenisoni</u> by the characters listed under that spe-
cies. Two populations can be distinguished by
differences in the morphology of their respective
adult males. Adults males collected from the
population surrounding South Georgia have supracaudal glands that are very similar to those of
the allopatric <u>P. tenisoni</u>. Adult males collected
from the Pacific population that is sympatric
with <u>P. tenisoni</u> have markedly smaller
supracaudal glands. Although the two populations
are almost certainly isolated from each other by
the waters of the Drake Passage, it is not known
whether they are similarly isolated in the eastern
hemisphere. These populations may eventually be
regarded as distinct species.

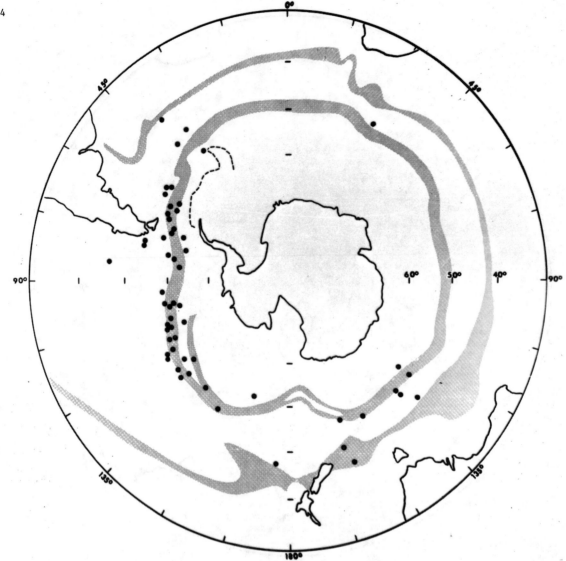

Fig. 6. Distribution of *Protomyctophum andriashevi*.

Protomyctophum (*P.*) *andriashevi* Becker, 1963
Fig. 6

Electrona (*P.*) *bolini* Fraser-Brunner, 1949, p.
  1100, partim (D 64).
Protomyctophum *andriashevi* Becker, 1963a, p. 19;
  1967a, p. 87 (near 42°S, 39°W).

Size. *Protomyctophum* *andriashevi* attains a
length of at least 50 mm. It has been collected
in the region of the Polar Front from 45°W west
to 120°E, extending north of the front in the
Australian, Chilean, and southwestern Atlantic
sectors. Except for a 27-mm specimen captured
near 60°S, 127°W (SOSC 354), individuals less
than 30 mm have been collected only north of the
polar front. The absence of records from most of
the Atlantic and Indian sectors may be due to
inadequate sampling.

Becker [1963a] reported *P.* *andriashevi* from
depths of 234 and 332 m at night and from 608 to
623 m during the day. *Eltanin* material includes
four individuals caught in trawls which did not
fish deeper than 100 m at night. The rest of the

specimens were caught in trawls which fished
deeper than 200 m.

Systematic notes. *Protomyctophum* *andriashevi*
is morphologically most similar to *P.* *normani*, *P.*
sp. B, and *P.* sp. C from which it can be distinguished by its low number of rakers on the first
gill arch (21 or less versus 23 or more) and the
morphology of the caudal luminous glands.

Protomyctophum (*P.*) sp. B
Fig. 7

Size. *Protomyctophum* sp. B attains a length of
about 70 mm. Caudal luminous glands develop in
males and females about 40 mm and 45 mm long,
respectively.

Distribution. The known distribution of *P.* sp.
B is shown in Figure 7. It has been collected in
and north of the region of the polar front west
from 40°W to 135°E. Large individuals with developed caudal glands were found, with one exception,
in the region of the polar front, and juveniles
were found between the polar front and about
50°S. The absence of records east of 40°W to

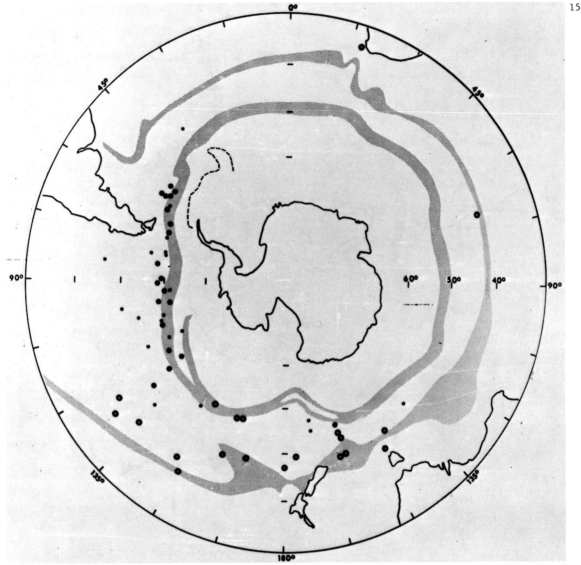

Fig. 7. Distributions of _Protomyctophum normani_ (black dots with white center) and _Protomyctophum_ sp. B (dots). Small dot indicates individual _Protomyctophum_ sp. B without developed caudal luminous glands, and large dot indicates individual _Protomyctophum_ sp. B with developed luminous glands.

135°E probably reflects inadequate sampling.

Protomyctophum sp. B has only been collected in trawls which fished deeper than 200 m.

Systematic notes. _Protomyctophum_ sp. B can be distinguished from the related species _P. normani_ and _P._ sp. C by its higher number of pectoral fin rays (usually more than 16 versus usually less than 16) and the presence of only a small supracaudal gland as well as a small (less than two photophore diameters in length) infracaudal gland in the male.

Protomyctophum (_P._) _normani_ (Tåning, 1932)
Fig. 7

Protomyctophum (_P._) _normani_ Andriashev, 1962, p. 231 (OB 394).

Size. _Protomyctophum normani_ attains a length of about 50 mm. Caudal luminous glands develop in males and females about 25 and 35 mm long, respectively.

Distribution. The distribution of the species I have assigned to _P. normani_ is shown in Figure 7. It does not include the type locality east of New Zealand because of the arbitrary assignmnt of Tåning's name to this form (see note below). _Protomyctophum normani_ has been collected in the Benguela Current, in the region of the subtropical convergence in the Indian sector, and in, as well as south of, the subtropical convergence region between 150°E and 120°W. It is absent from the waters between 120°W and Chile. Captures closest to the Polar Front are in the vicinity of the southeastward flow of water in the Pacific subantarctic region.

The study material includes individuals captured in trawls which did not fish deeper than 100 m.

Systematic notes. Two forms similar to the holotype of _P. normani_ are present in the study

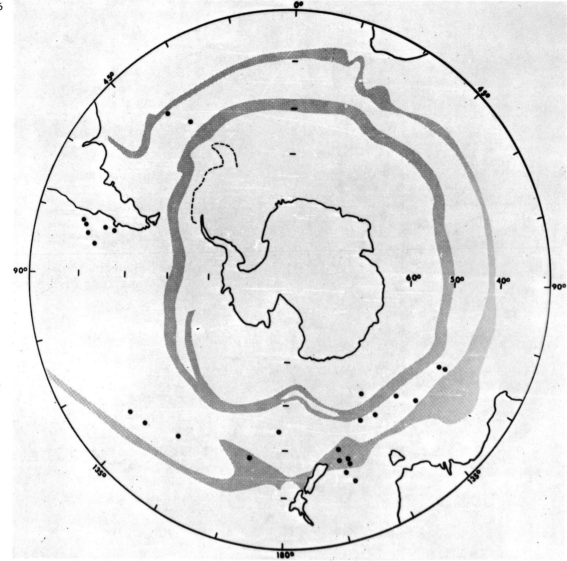

Fig. 8. Distribution of _Protomyctophum_ sp. C.

material. I have assigned Tåning's name to this form only because it has been collected closer to the type locality than P. sp. C. The holotype of P. _normani_ is a 21-mm juvenile that lacks caudal luminous glands, and the two forms can be distinguished in the western Pacific sector only after the caudal glands have developed. Adult P. _normani_ can be distinguished from P. sp. C by the absence in males and females of infracaudal and supracaudal glands, respectively.

Protomyctophum P. sp. C
Fig. 8

Protomyctophum _normani_ Becker, 1963a, p. 17; 1967a, p. 89 (near 42°S, 39°W).

Size. _Protomyctophum_ sp. C attains a length of about 65 mm. Caudal luminous glands develop in males and females about 40 and 50 mm long, respectively. Gravid females were found in a sample collected in December.

Distribution. The known distribution of P. sp. C is shown in Figure 8. _Protomyctophum_ sp. C occurs in subantarctic and subtropical convergence

waters near 45°W, east from 120°E to 135°W, and between 40°S and 50°S off Chile. It is apparently absent from between 125°W and 85°W. On the other hand, its absence in most of the Atlantic and Indian sectors may be due to inadequate sampling.

Systematic notes. _Protomyctophum_ sp. C is distinguished from P. sp. B and P. _normani_ by the presence of supracaudal and infracaudal luminous glands in both sexes, the latter gland extending one half the distance between the procurrent rays and the anal fin in the male. Individuals caught near South America have fewer gill rakers (usually less than 24 versus usually more than 24) and fewer AO photophores (15 versus 16-17) than individuals collected near New Zealand.

Protomyctophum (P.) bolini
(Fraser-Brunner, 1949)
Fig. 9

Myctophum _tenisoni_ Norman, 1930, p. 321 (part, D 72).--Norman, 1937, p. 84 (part, BANZARE 69).
Electrona (P.) _bolini_ Fraser-Brunner, 1949, p. 1100 (part, D 72, 671, 849).
Protomyctophum (P.) _bolini_ Andriashev, 1962, p.

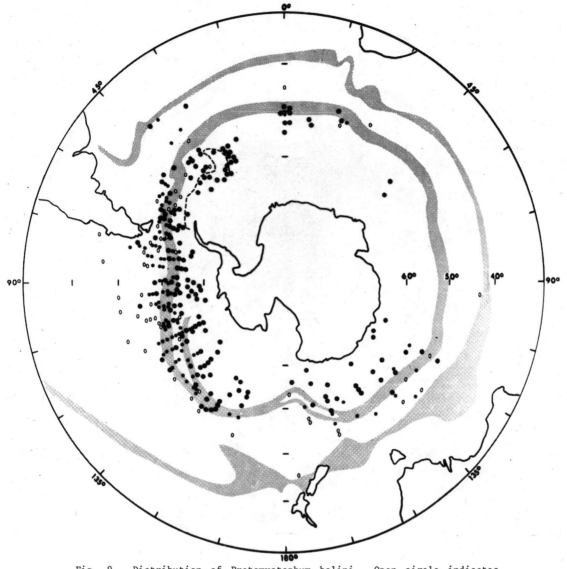

Fig. 9. Distribution of Protomyctophum bolini. Open circle indicates individuals less than 30 mm long, large dot indicates individuals more than 29 mm long, and small dot indicates individuals of both size ranges in same sample.

232 (OB 214, 368, 409, 413, 455, 464, 468; Slava 57, 73, 77, 96, 97, 100, 101). --Becker, 1963a, p. 16; 1967a, p. 90 (near 42°S, 39°W).

Size. Protomyctophum bolini grows to a length of about 60 mm. Gravid females about 52 mm long were caught during November to February, April, and June.

Distribution. Figure 9 shows the known distribution of P. bolini. The type locality is not included in the figure because the position reported by Fraser-Brunner [1949] does not correspond to the position listed in the Discovery reports for station 1559. Protomyctophum bolini is distributed in a circumpolar band generally centered in the Polar Front. With few exceptions, juveniles less than 30 mm long have been collected in and north of the Polar Front region, whereas larger individuals have been taken mainly in and south of the Polar Front. Some large specimens have been caught north of the Polar Front south of Australia, between 105°W and Chile, and east

of the Brazil Current region. The southern limits of its distribution are nearly the same as those of P. anderssoni, roughly corresponding to the region of the Antarctic divergence. However, P. bolini is also found south of the Weddell-Scotia confluence, where P. anderssoni is absent. The lack of records in much of the Indian sector may be due to inadequate sampling.

Hierops Fraser-Brunner, 1949

The subgenus Hierops includes seven or eight species, four of which are known from the northern hemisphere. Protomyctophum (H.) arcticum (Lütken, 1892) is known only from the North Atlantic between 45°N and 70°N [Bolin, 1959; Becker, 1967a] and P. (Hierops) thompsoni (Chapman, 1944) is known only from the subarctic North Pacific [Becker, 1963b]. [Becker, 1963b] recognized allopatric eastern and western populations of an additional species, P. (H.) crockeri (Bolin, 1939), in transitional waters of the North Pacific

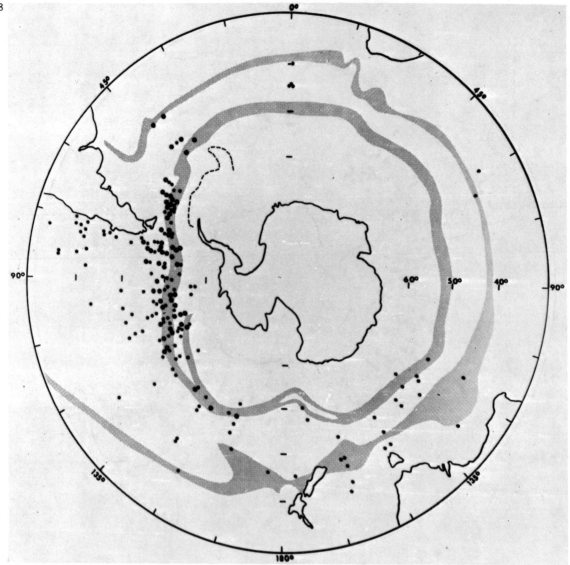

Fig. 10. Distribution of Protomyctophum (Hierops) parallelum. Small dot indicates individual less than 40 mm long. Large dot indicates that sample includes individuals larger than 39 mm.

and described some specimens from near Hawaii which he tentatively assigned to this species. Wisner [1971] placed the latter specimens in a new species, P. (H.) beckeri, known from five localities between 7°N and 20°N and 144°W and 173°W. Wisner also placed what had previously been considered a Chilean population of P. (Hierops) crockeri in a new species, P. (Hierops) chilensis. Four species of P. (Hierops) are represented in the study collections.

Protomyctophum (Hierops) parallelum
(Lönnberg, 1905)
Fig. 10

Myctophum parallelum Lönnberg, 1905, p. 62 (northwest of South Georgia).
Protomyctophum (Hierops) parallelum Andriashev, 1962, p. 238 (OB 416, 417).--Nafpaktitis and Nafpaktitis, 1969, p. 8 (AB 7112, 7351).--Craddock and Mead, 1970, p. 30 (B13 41).

Size. Protomyctophum parallelum attains a length of about 50 mm. Gravid females as small as 32 mm were taken during April, July, August, and October.
Distribution. Figure 10 shows the known distribution of P. parallelum. This species has been collected in a circumpolar band which extends from the Polar Front to about 40°S in the areas sampled by the Eltanin; it has not been found near the Polar Front in the remaining longitudes. Individuals larger than 40 mm have been collected only near the Polar Front between 45°W and 165°W. The northern limits of distribution roughly correspond to the region of the subtropical convergence. Three juveniles less than 30 mm long were captured in three trawls which did not fish deeper than 150 m, and nearly 400 juveniles were taken in a single trawl which did not exceed 366 mm. The remainder of the material was collected in trawls which fished deeper than 450 m, which indicates that P. parallelum is a relatively deep-living species.

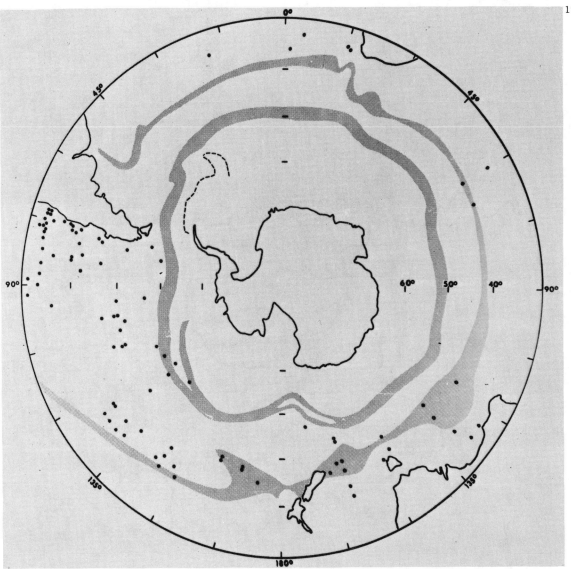

Fig. 11. Distribution of Protomyctophum (Hierops) subparallelum.

Systematic notes. Protomyctophum (H.) parallelum is easily distinguished from its congeners by
a number of characters, particularly the wide sep-
aration of its Prc photophores.

Protomyctophum (Hierops) subparallelum
(Tåning, 1932)
Fig. 11

Myctophum parallelum Norman, 1930, p. 320 (D 78,
  85, 100, 100c, 101).
Myctophum arcticum subparallelum Tåning, 1932,
  p. 128 (east of New Zealand).
Protomyctophum (H.) subparallelum Andriashev, 1962
  p. 235 (OB 351, 393, 394, 419, 436, 441, 442,
  444, 446).--Nafpaktitis and Nafpaktitis, 1969,
  p. 8 (AB 7112, 7133, 7351).--Craddock and Mead,
  1970, p. 30 (B13 35 stations 30°-34°S, 72°-
  92°W).

Size. Protomyctophum subparallelum attains a
length of about 30 mm. Gravid females as small
as 27 mm have been taken in May, July, September,
and October.
Distribution. The known distribution of P. sub-

parallelum is shown in Figure 11. This species
has been taken in and north of the region of the
subtropical convergence in the eastern Atlantic,
Indian, and Australian sectors and farther south
in the Pacific sector. It is apparently absent
in the waters of the Brazil and Falkland currents.
Protomyctophum subparallelum was collected by
the Eltanin only in trawls which fished deeper
than 275 m. Nafpaktitis and Nafpaktitis [1969]
and Craddock and Mead [1970] report specimens
from two trawls which did not exceed 100 m and
150 m, respectively.
Systematic notes. Protomyctophum subparallelum
is morphologically very similar to P. arcticum
and is presumably very closely related to that
species. Both species are related, to a lesser
degree, to P. thompsoni.

Protomyctophum (Hierops) sp. D
Fig. 12

Size. Protomyctophum sp. D attains a length of
about 35 mm.
Distribution. The known distribution of P. sp.

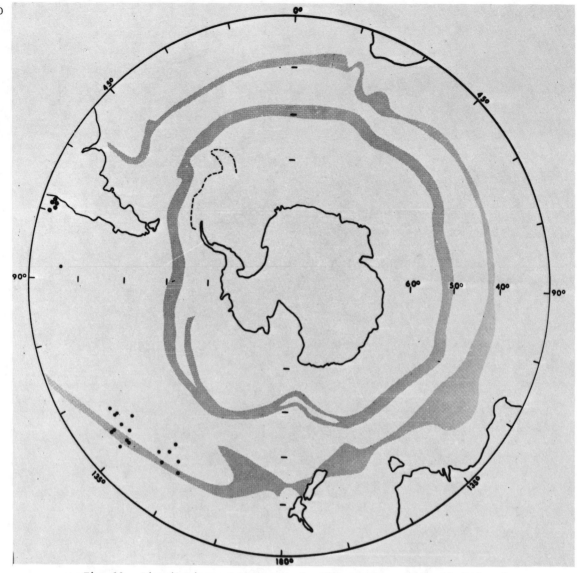

Fig. 12. Distributions of Protomyctophum (Hierops) sp. D (small dot) and
of Protomyctophum (Hierops) crockeri south of 30°S (large dot).

D is shown in Figure 12. This form has been collected in the region of the subtropical convergence from near 150°W east to 87°W. The southernmost records correspond to the northernmost
records of P. subparallelum.

Systematic notes. Protomyctophum sp. D and the
very similar, and presumably most closely related,
species P. beckeri are distinguished from other
Hierops species by a combination of characters:
the juxtaposition of their Prc photophores, the
relative straightness of their SAO photophores,
and the lower number of rakers on the first gill
arch (less than 20 versus more than 20). Protomyctophum sp. D is distinguished from P. beckeri by
its higher number of rakers on the first gill
arch, its narrower interorbital space, and its
more slender body.

Protomyctophum (Hierops) chilensis
(Wisner, 1971)
Fig. 12

Protomyctophum crockeri Bussing, 1965, p. 200
(ELT 742).

Protomyctophum chilensis Wisner, 1971, p. 39
(five localities near the Chilean coast).

Size. Protomyctophum chilensis attains a length
of at least 37 mm off Chile.

Distribution. Figure 12 shows the known distribution of P. chilensis south of 30°S. This
species has also been collected near 22°S, 79°W
in the Peru Current [Wisner, 1971] (see systematic
notes).

Systematic notes. Wisner [1971] recognized the
Chilean population as specifically distinct from
P. crockeri on the basis of a larger head length.

Electrona Goode and Bean, 1896

Paxton [1972] placed Metelectrona Wisner, 1963
in synonymy with Electrona. Although Wisner
[1963] did not provide a definitive diagnosis for
Metelectrona, Paxton points out that with the
exception of elevated VO2 photophores the remaining characters emphasized by Wisner appear to be
a mosaic in the various species of Electrona.

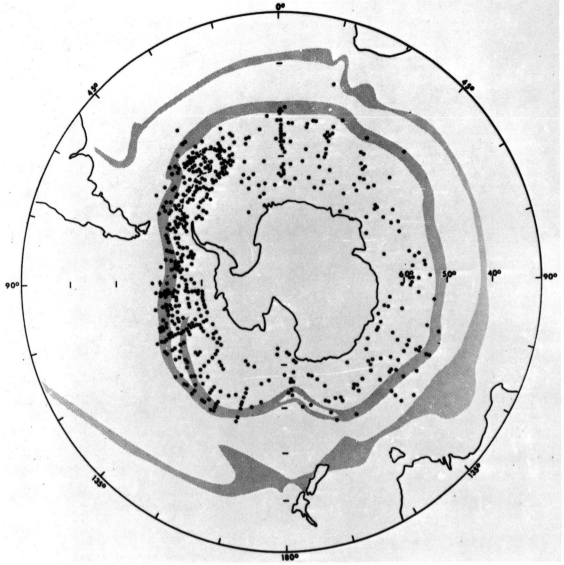

Fig. 13. Distribution of *Electrona antarctica*.

Paxton regards the elevated VO2 photophores as specifically, but not generically, significant. Moser and Ahlstrom [1974], however, state that an inherently unique larval morphology, particularly with respect to the shape of the gut, pigmentation pattern, formation of the dorsal fin, and relative position of the anus and anal fin, strongly suggests recognition of *Metelectrona* as a valid genus. Although I follow their suggestion, it seems apparent from examination of a large number of adults of all known species of both genera that *Electrona* and *Metelectrona* are closely related. *Electrona*, as recognized here, includes five species, all of which are present in the study material. *Electrona rissoi* is the only member of the genus found north of the 'Southern' Ocean.

Electrona antarctica (Günther, 1878)
Fig. 13

Scopelus antarcticus Günther, 1878, p. 184 (Challenger 156, 157).
Myctophum antarcticum Regan, 1913, p. 234 (near Coates Land).--Norman, 1930, p. 332., (part D 116, 121, 197, 202); 1937, p. 85 (part, BANZARE 27, 33, 45, 96).

Size. *E. antarctica* attains a length of about 100 mm. Gravid females about 61 mm long were found in samples taken in all months except June, July, and November.

Distribution. The distribution of *E. antarctica* is shown in Figure 13. Additional records are avaiable from the literature. Andriashev [1962] and Norman [1930] reviewed the synonymy of several species of *Electrona* and placed *Myctophum antarcticum* of Gilbert [1911], Regan [1914], and Waite [1916] in the synonymy of *E. subaspera*. I have examined Regan's specimen, and it is *E. subaspera*. The descriptions offered by Gilbert and Waite, particularly regarding the elevation of the PO 5 photophore, warrant the exclusion of their material from the synonymy of *E. antarctica*. *Myctophum antarcticum*, as reported by Brauer [1906], includes at least one individual of another species; the elevated AO photophores evident in his Figure 82c implicates *E. paucirastra* or *Metelectrona*. Although Norman [1930] and Andriashev [1962] have

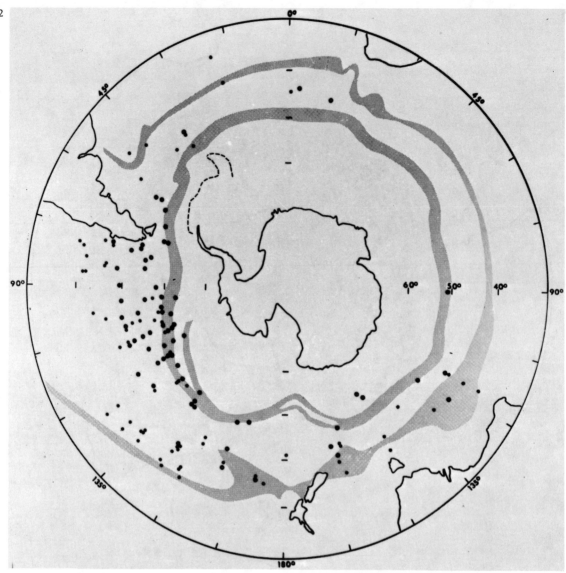

Fig. 14. Distribution of _Electrona subaspera_. Small dot indicates individuals less than 40 mm long, and large dot indicates individuals larger than 39 mm long.

accepted the specimen reported by Pappenheim [1914] from northwest of Prince Edward Island as _E. antarctica_, the unusual locality and inadequate description make the indentification doubtful. I examined the specimen reported from _Discovery_ collection 267 by Norman [1930] and found it to be _Metelectrona ventralis_; the remainder of Norman's material that was available for study proved to be _E. antarctica_. Four of the ten BANZARE records reported by Norman [1937] proved to be _E. antarctica_. Additional records of _E. antarctica_ were not included in Figure 13 because they would not significantly alter the distributional pattern evident in the figure. These additional records include the six remaining BANZARE stations, the material of Lönnberg [1905], Pappenheim [1912], Krefft [1958], and Dewitt and Taylor [1960], and the extensive collections reported by Andriashev [1962].

Electrona antarctica has been collected south of the Polar Front in all longitudes. It has also been found in 18 _Eltanin_ and _Discovery_ samples north of the Polar Front, 11 of which are less than 1° of latitude beyond the northern limit of the front. Five were _Eltanin_ samples collected from 1°-5° north of the front in the southwest Atlantic sector, near 90°W, and south of Australia, and two were _Discovery_ records from near 15°E. However, the data with one of the _Discovery_ samples, (D 407 near 35°S) did not correlate with published station data, and it is not included in Figure 13. With the exception of a 60-mm specimen taken very closely to the Polar Front (Elt 2291), all individuals captured in subantarctic waters were less than about 50 mm long. All individuals captured south of the Weddell-Scotia confluence between 35°W and 55°W were more than about 40 mm long.

Electrona antarctica apparently undergoes a diel vertical migration during at least part of the year. Andriashev [1962] reports the capture of a single specimen in the surface at night. At 33

23

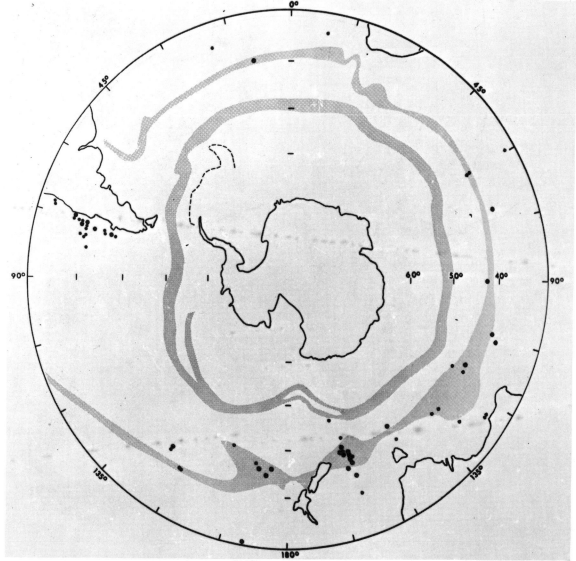

Fig. 15. Distribution of _Electrona paucirastra_. Small dot indicates
individuals less than 28 mm long, and large dot indicates individuals
larger than 27 mm long.

Discovery stations this species was captured above
100 m at night; 4 samples came from the upper 5
m. Electrona antarctica was found in 18 Eltanin
trawls which did not exceed 100 m. The two shallowest of the 21 daytime Discovery records were
from 130 m and 155 m; the shallowest Eltanin daytime captures were at 275 m and 44 m. The latter
record (Elt 1942) is from shallow water (256 m).

Electrona subaspera (Günther, 1864)
Fig. 14

Scopelus (Dasyscopelus) subaspera Günther, 1864,
    p. 411 (near 43°S, 123°E).
Myctophum antarcticum Regan, 1914, p. 1 (near
    55°S, 120°W).
Myctophum subasperum Norman, 1930, p. 323, (part
    D 104).
Electrona subaspera Andriashev, 1962, p. 245 (OB
    72, 389, 394, 395, 397, 398, 411, 415, 416,
    417, 419).

Size. Electrona subaspera attains a length of
about 120 mm. A 102 mm gravid female was collected in July (Elt 1730).

Distribution. With the exception of the Scopelus subaspera reported by Lütken [1892] from the
latitude of New York, additional records from the
literature included in the synonymy of E. subaspera by Norman [1930] and Andriashev [1962] do
not alter the distributional pattern evident in
Figure 14. The synonymy, however, is in need of
revision. The material reported by Norman [1930,
1937] included E. carlsbergi, E. paucirastra, and
M. ventralis. Some additional reports, including
the New York record, are undoubtedly in error.

Figure 14 shows the distribution of E. subaspera. This species has been collected primarily
between the Polar Front and the subtropical convergence. A juvenile collected with M. ventralis
near 20°W (D 78) is the only record from farther
north. Two large individuals taken near 150°E in
February, when surface temperatures were great-

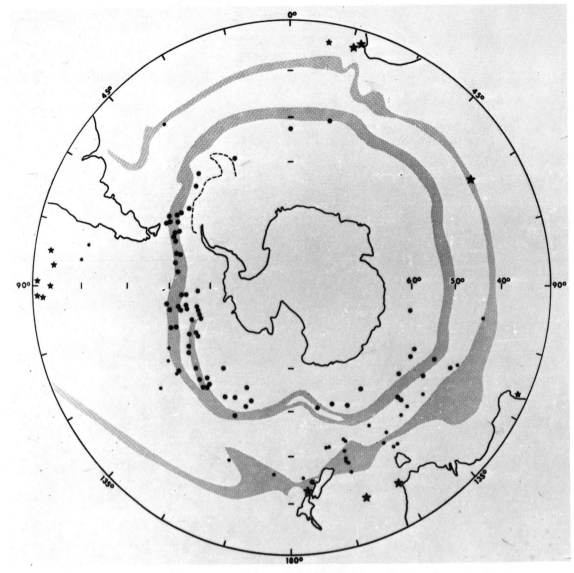

Fig. 16. Distribution of *Electrona carlsbergi* (small dot indicates individuals less than 50 mm long and large dot indicates individuals larger than 49 mm long) and of *Electrona rissoi* south of 30°S (small star indicates individuals less than 30 mm long and large star indicates individuals larger than 50 mm long.

er than 2.4°C, are the only records from south of the Polar Front. Juveniles, less than about 40 mm long, have been collected farther north than have larger individuals. The northern limits of distribution overlap the southern limits of distribution of *E. paucirastra*, *M. ventralis*, and *M. sp. A* and correspond to the subtropical convergence, except east of 120°W in the Pacific sector. The southern limits of distribution overlap the northern limits of distribution of *E. antarctica* in the waters of the Polar Front. The absence of records between 15°E and 90°E may reflect inadequate sampling. *Electrona subaspera* has frequently been captured on the surface at night.

Electrona paucirastra Bolin (Andriashev, 1962)
Fig. 15

Myctophum subasperum Norman, 1930, p. 323 (part D 78); 1937, p. 85 (BANZARE 72).

Electrona paucirastra Andriashev, 1962, p. 280 (near 40°S, 71°E). Bussing, 1965, p. 200 (ELT 743).--Nafpaktitis and Nafpaktitis, 1969, p. 9 (AB 7094, 7127, 7133).

Size. *Electrona paucirastra* attains a length of about 60 mm. Gravid females about 60 mm long were collected in August, September, and October.

Distribution. The known distribution of *E. paucirastra* is shown in Figure 15. It has been collected in the region of the subtropical convergence near and west of New Zealand and between 35°S and 50°S near Chile. the northernmost record, near 170°W, is that of a specimen transferred to the British Musseum from the Godeffroy Museum in 1878. The locality data with the specimen are 30°S, 170°W and may only be approximate. Juveniles less than about 30 mm long have beem taken farther north and farther south than have larger individuals. *Electrona paucirastra* appears

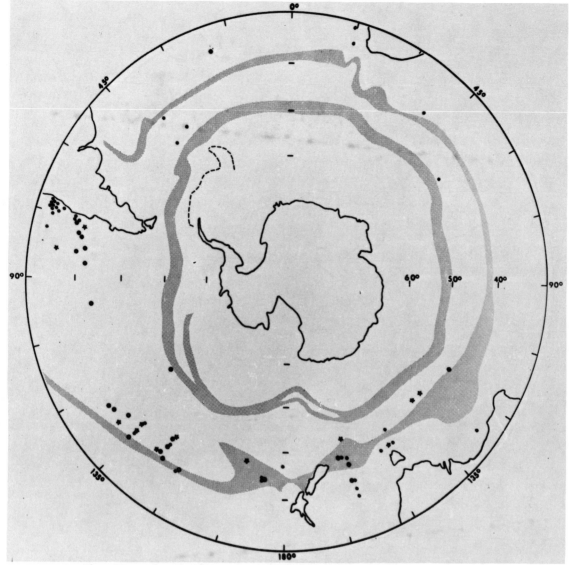

Fig. 17. Distributions of _Metelectrona ventralis_ south of 30°S (small dot) and of _Metelectrona_ sp. A (large dot). Unidentifiable specimens of _Metelectrona_ are represented by a star.

to be absent from the southwestern Atlantic and central-eastern Pacific sectors. Off Chile its distribution overlaps those of M. ventralis to the north, M. sp. A to the west, and E. subaspera to the north. It is the only species in the genus found between 40°S and 50°S off Chile.

Electrona paucirastra has been captured on the surface at night.

Systematic notes. Electrona paucirastra is most similar to E. subaspera [Moser and Ahlstrom, 1970, 1974]. It also is quite similar to Metelectrona. Individuals from off Chile appear to have a more slender body and somewhat longer male supracaudal luminous gland than have individuals from the allopatric population farther west.

Electrona carlsbergi (Tåning, 1932)
Fig. 16

Myctophum subasperum Norman, 1930, p. 85 (part, BANZARE 69).
Myctophum carlsbergi Tåning, 1932, p. 126 (near 44°S, 173°E).

Electrona carlsbergi Andriashev, 1962, p. 243 (OB 394, 409, 455, 464; SLAVA 96, 100).--Becker, 1963a, p. 22; 1967a, p. 91 (near 42°S, 39°W).

Size. E. carlsbergi attains a length of at least 90 mm.

Distribution. Figure 16 shows the known distribution of E. carlsbergi. This species has been collected primarily south of the subtropical convergence. Inadequate sampling probably accounts for the absence of records in the Indian sector, whereas the species appears to be absent from waters near the Polar Front between 30°W and 60°W. Except for two records south of Australia and the type locality near New Zealand, individuals larger than 50 mm have been taken in or near the Polar Front. Smaller individuals, except for two records near 110°W, have been collected north of the Polar Front.

Electrona carlsbergi larger than 40 mm apparently occur below 200 m, whereas smaller individuals have been taken at depths as shallow as 60 m.

26

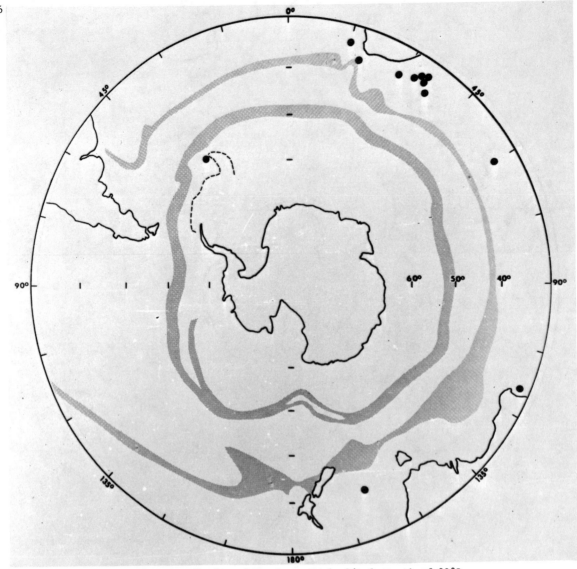

Fig. 18. Distribution of Benthosema suborbitale south of 30°S.

Electrona rissoi (Cocco, 1829)
Fig. 16

Electrona rissoi McCulloch, 1915, p. 104 (off SE
    Australia).--Andriashev, 1962, p. 249 (Cook
    Strait, New Zealand).--Trunov, 1968 (near 32°S,
    16°E).--Nafpaktitis and Nafpaktitis, 1969, p.
    10 (AB 7133).--Craddock and Mead, 1970, p. 27
    (B13 17, 20, 23, 26, 28, 30, 31).
Myctophum rissoi Norman, 1930, p. 320 (D 87, 101)

    Size. E. rissoi attains a length of at least
70 mm.
    Distribution. Figure 16 shows the distribution
of E. rissoi south of 30°S. This species is also
known from north of 30°N and east of 40°W in the
northeastern Atlantic Ocean [Bolin, 1959; Backus
et al., 1970], from the equatorial Atlantic Ocean
between 45°W and 0° and near 10°N, 25°W [Becker,
1967a; Blauche and Stauch, 1964a; Backus et al.
1970], from the Benguela Current near 28°S, 14°E
[Trunov, 1968], from the equatorial Indian Ocean
[Nafpaktitis and Nafpaktitis, 1969], from the
northeastern Pacific between 20°N and 0°N east of

155°W [Andriashev, 1962; Moser and Ahlstrom,
1970], and from Japan (T. Uyeno, personal communication, 1972). One or two types of larvae re-
sembling those of E. rissoi have been reported
from two latitudinal bands in the eastern tropical
Pacific [Ahlstrom, 1971]. The region of the subtropical convergence marks the southernmost limits
of E. rissoi.
    Juvenile E. rissoi have been captured in trawls
which did not fish deeper than 175 m, whereas
adults appear to be restricted to deeper layers.
    Systematic notes. Electrona rissoi is a
specialized deep-bodied and foreshortened form
[Paxton, 1972]. Electrona rissoi larvae share a
number of characters with the larvae of E. carlsbergi [Moser and Ahlstrom, 1974].

Metelectrona Wisner, 1963.

    As mentioned above, Paxton [1972] synonimizes
Metelectrona with Electrona on the basis of comparative adult morphology, whereas Moser and Ah-
strom [1974] suggest that comparative larval morphology warrants its distinction as a valid genus.

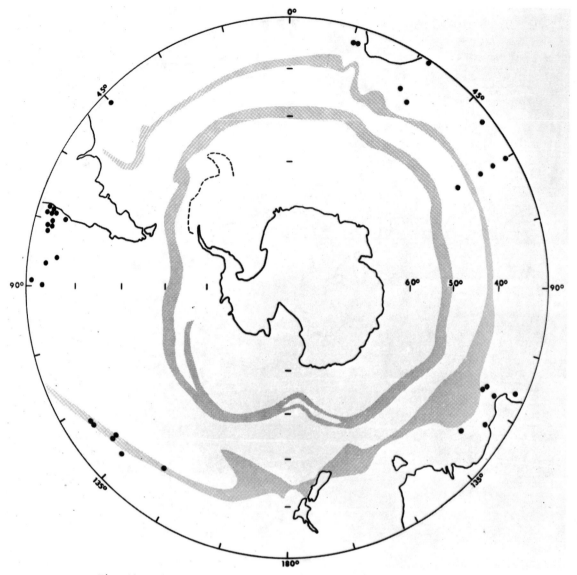

Fig. 19. Distributions of <u>Diogenichthys atlanticus</u> south of 30°S.

The latter authors also seem to suggest that <u>Metelectrona</u> shows some affinity to the genus <u>Hygophum</u>. These conflicting interpretations are not surprising, however, in view of the frequently mosaic nature of evolution [Mayr, 1963, p. 596]. The two known species of <u>Metelectrona</u> are present in the study material; one is undescribed.

<u>Metelectrona ventralis</u> (Becker, 1963)
Fig. 17

<u>Myctophum antarcticum</u> Norman, 1930, p. 322 (part, D. 267).
<u>Electrona ventralis</u> Becker, 1963a, p. 26; 1967a; p. 91 (near 42°S, 39°W).
<u>Metelectrona ahlstromi</u> Wisner, 1963, p. 25 (near 42°S, 179°W).--Bussing, 1965, p. 200 (ELT 190, 742, 743). --Craddock and Mead, 1970, p. 30. (B 13, 4, 5, 44, 47, 48, 49, 50, 51, 52, 53, 59, 63).

<u>Size</u>. <u>Electrona ventralis</u> attains a length of at least 60 mm.
<u>Distribution</u>. The known distribution of <u>M. ventralis</u> south of 30°S is shown in Figure 17.

This species is also known from one locality farther north, near 24°S in the Benguela Current (D 267). It has been collected between 33°S and 40°S off Chile and in or near the region of the subtropical convergence in the Atlantic, Indian, and western Pacific sectors. It is absent from the central-eastern Pacific sector.

<u>Metelectrona ventralis</u> has been collected in several trawls that did not fish deeper than 100 m, including one that did not fish deeper than 5 m.

<u>Systematic notes</u>. <u>Metelectrona ventralis</u> is very closely related to <u>M</u>. sp. A, from which it can be distinguished only by a more anterior position of the Vn photophore and the presence of single, rather than double, infracaudal and supracaudal luminous glands. Damaged juveniles that lack a Vn photophore and supracaudal glands are plotted as <u>M</u>. sp.

<u>Metelectrona</u> sp. A
Fig. 17

<u>Myctophum subasperum</u> Norman, 1930, p. 323 (part D 78).

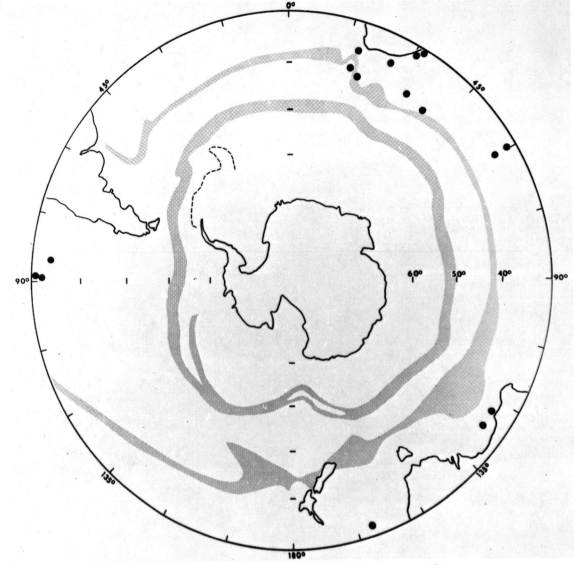

Fig. 20. Distribution of <u>Hygophum hygomi</u> south of 30°S.

Size. <u>Metelectrona</u> sp. A attains a length of
at least 60 mm.
   <u>Distribution</u>. Figure 17 shows the known distribution of <u>M</u>. sp. A. This species has been collected in and south of the subtropical convergence
between Chile and 115°E and near 35°S in the At-
lantic sector. Unlike <u>E</u>. <u>paucirastra</u> and <u>M</u>. <u>ven-
tralis</u>, it has been taken in oceanic waters be-
tween 125°W and 85°W, including one locality near
the Polar Front at approximately 120°W. Individuals more than 27 mm long have been collected
only in the Tasman Sea and near the Chatham Islands.
   <u>Metelectrona</u> sp. A was found in several trawls
which did not fish deeper than 100 m.

Benthosema Goode and Bean 1896

   The genus <u>Benthosema</u>, which includes five or
six species, is related to <u>Diogenichthys</u> [Moser
and Ahlstrom, 1970]. <u>Benthosema pterotum</u> (Alcock,
1891), <u>B</u>. <u>fibulatum</u> (Gilbert and Cramer, 1897),
and <u>B</u>. <u>panananense</u> (Tåning, 1932) are distributed
in low latitudes. <u>Benthosema glaciale</u> (Rhein-

hardt, 1837) is known from the North Atlantic and
may occur in waters of the Arctic Basin [Bolin,
1959]. Becker [1967b] suggests that <u>b</u>. <u>suborbi-
tale</u>, as recognized by Bolin [1959], could include
more than one species.

<u>Benthosema suborbitale</u> (Gilbert, 1913)
Fig. 18

<u>Benthosema suborbitale</u> Nafpaktitis and Nafpakti-
   tis, 1969, p. 11 (AB 7112).

   Size. <u>Benthosema suborbitale</u> attains a length
of about 30 mm. Nafpaktitis and Nafpaktitis [1969]
found gravid females as small as 24 mm.
   <u>Distribution</u>. Figure 18 shows the known distribution of <u>B</u>. <u>suborbitale</u> south of 30°S. The
<u>B</u>. <u>fibulatum</u> reported by Norman [1930] from D 257,
near 35°S, 10°E, was examined and found to be <u>B</u>.
<u>suborbitale</u>. However, it is not included in
Figure 18 because of a discrepancy between Norman's station data and the data published in the
station list. This species is also known from
the North Atlantic between the equator and 40°N

[Bolin, 1959; Backus et al., 1965, 1969, 1970; Becker, 1967a; Discovery data], from the Indian Ocean between the equator and 35°S [Nafpaktitis and Nafpaktitis, 1969; Discovery data], from the northwestern Pacific Ocean near Japan [Gilbert, 1913], and from the central-eastern tropical Pacific Ocean between 20°N and 30°N and between 10°S and 21°S [Berry and Perkins, 1966; R. L. Wisner, unpublished data, 1973]. Becker [1978b] reports the existence of antitropical populations in the subtropical Pacific Ocean. The study area includes the southern limits of the species. The southernmost record, near 34°W, evidently prompted Bolin [1959] to state that this species may occur beyond 50°, where warm water moves poleward. Yet this specimen, the only one known from south of 40°S, was actually captured in an area where the predominant flow is equatorward and cold. The lack of additional records from the southwestern Atlantic sector introduces the possiblity that the specimen was mislabeled.

Benthosema suborbitale has been collected at depths less than 100 m.

Diogenichthys Bolin, 1939

The genus Diogenichthys includes three species. Diogenichthys laternatus (Garman, 1899) and D. panurgus Bolin, 1946, are closely related and are distributed in equatorial waters of the Pacific and Indian oceans, respectively [Nafpaktitis and Nafpaktitis, 1969]. Diogenichthys atlanticus is present in the study material.

Diogenichythys atlanticus (Tåning, 1928)
Fig. 19

Myctophum laternatum Norman, 1930, p. 324 (part D 100, 285, 287, 288, 289, 296).
Diogenichthys atlanticus Bussing, 1965, p. 202 (Elt 80, 742).--Nafpaktitis and Nafpaktitis, 1969, p. 13 (AB 7094, 7096, 7100, 7112, 7123).--Craddock and Mead, 1970, p. 27 (B 13; 18 stations 30°-35°S, 70°-90°W).

Size. Diogenichthys atlanticus attains a length of about 25 mm. Gravid females as small as 22 mm were found in the study material.

Distribution. The known distribution of D. atlanticus south of 30°S is shown in Figure 19. It is also known from the western North Atlantic Ocean between 20°N and 40°N [Backus et al., 1969], across the North Atlantic between 30°N and 40°N [Bolin, 1959], in the equatorial Atlantic east of 35°W between 15°N and 5°S, and southward along the west African coast to 18°S [Backus et al., 1965; Becker, 1967a; Blache and Stauch, 1965b; Discovery data]. It has been reported from the Indian Ocean as far north as 22°S [Nafpaktitis and Nafpaktitis, 1969], from the southeastern Pacific Ocean as far north as 20°S near the South American coast [Bussing, 1965], and 10°S offshore [Ahlstrom, 1971], and from the northeastern Pacific between 20°N and 40°N [Moser and Ahlstrom, 1970]. Ahlstrom [1971] reports a single specimen from the eastern equatorial Pacific where the endemic D. laternatus is very abundant. Becker [1967b] reports on a form related to D. atlanticus from the northwestern Pacific Ocean. The southern limits of distribution of this species generally correspond with the

region of the subtropical convergence. It appears to be absent from waters near New Zealand.

Diogenichthys atlanticus has frequently been collected at depths less than 100 m.

Hygophum Bolin, 1939

The genus Hygophum includes about nine species [Becker, 1965; Wisner, 1971]. It is related to Myctophum and Symbolophorus [Paxton, 1972; Moser and Ahlstrom, 1970], Hygophum benoiti (Cocco, 1838) is confined to the North Atlantic subtropical gyre, and H. reinhardti (Lütken, 1892) is distributed in subtropical waters of both hemispheres [Becker, 1965]. One or two of the four or five species that are present in the study area belong to the H. macrochir species-group, in which Becker [1965] recognizes four tropical and subtropical species.

Hygophum hygomi (Lütken, 1892)
Fig. 20

Scopelus hygomii Lütken, 1892, p. 256 (part near 35°S, 26°E; see Bolin, 1959, p. 7).
Myctophum hygomii Waite, 1904, p. 153 (Lord Howe Island).
Hygophum hygomi Nafpaktitis and Nafpaktitis, 1969, p. 19 (AB 7100, 7112).--Craddock and Mead, 1970, p. 28 (B 13 19, 29, 30).

Size. The morphology of early developmental stages of H. hygomi indicates a possible relationship with H. benoiti [Becker, 1965]. This species attains a length of at least 60 mm.

Distribution. Figure 20 shows the distribution of H. hygomi south of 30°S. Bolin [1959] discusses additional material from North Island, New Zealand, but does not record the specific locality. The species is also known from the North Atlantic Ocean between 20°N and 46°N and the Indian Ocean as far north as 21°S [Bolin, 1959; Becker, 1965; Backus et al., 1969; Nafpaktitis and Nafpaktitis, 1969]. It is apparently absent in the near-shore waters of Chile. Its southern limits of distribution generally correspond to the region of the subtropical convergence.

Hygophum hygomi has been collected in the surface [Craddock and Mead, 1970].

Hygophum hanseni (Tåning, 1932)
Fig. 21

Scopelus hygomii Lütken, 1892, p. 256 (part, near 40°S, 41°E; see Bolin, 1959, p. 7)
Myctophum benoiti var. reinhardti Barnard, 1925, p. 242 (Cape Point to Agulhas Bank; see Tåning, 1932, p. 133).
Myctophum macrochir Norman, 1930, p. 326 (part D 71).
Myctophum hanseni Tåning, 1932, p. 132 (near 42°S, 175°E).
Hygophum hanseni Becker, 1965, p. 95 (near 42°S, 159°E and 42°S, 39°W).--Nafpaktitis and Nafpaktitis, 1969, p. 19 (AB 7324).--Craddock and Mead, 1970, p. 27 (part, B13 17, 41).

Size. Hygophum hanseni attains a length of at least 45 mm.

Distribution. The known distribution of H. hanseni is shown in Figure 21. It has been col-

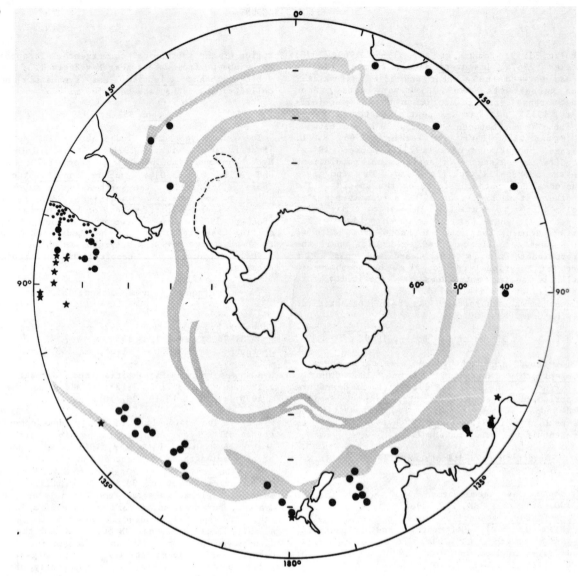

Fig. 21. Distributions of Hygophum hanseni (large dot) and Hygophum bruuni (small dot) and of Hygophum macrochir group south of 30°S (star).

lected in the region of the subtropical convergence in most sectors of the Atlantic, Pacific,
and Indian oceans and farther south in the Pacific
sector. It is generally excluded from the near-
shore waters of Chile.

Hygophum hanseni has been collected at depths
less than 100 m.

Systematic notes. Hygophum hanseni differs
from the closely related species H. bruuni in a
number of characteristics, including a lower
number of gill rakers on the first gill arch (18
or less versus 19 or more) and a shorter male
supracaudal luminous gland.

Hygophum bruuni Wisner, 1971
Fig. 21

Hygophum hanseni Bussing, 1965, p. 202 [Elt 190,
742, 743].--Craddock and Mead, 1970, p. 27
(part; see Appendix 1).
Hygophum bruuni Wisner, 1971, p. 41 (30°-34°S
near Chile).

Size. Hygophum bruuni attains a length of about
50 mm.

Distribution. The known distribution of H.
bruuni is shown in Figure 21. This species is
restricted to between 30°S and 45°S off Chile.
H. bruuni has been taken in surface waters.

Hygophum macrochir group
Fig. 21

Hygophum hanseni Craddock and Mead, 1970, p. 27
(part; see Appendix 1).

The distribution of specimens belonging to the
H. macrochir group south of 30°S is shown in
Figure 21.

Becker [1965] distingishes four species in the
H. macrochir group: H. taaningi Becker, 1965, in
the central gyre of the North Atlantic; H. macrochir (Günther, 1864), in the equatorial Atlantic;
H. proximum Becker, 1965, in the tropical Indian
and Pacific oceans; and H. atratum (Garman, 1899)

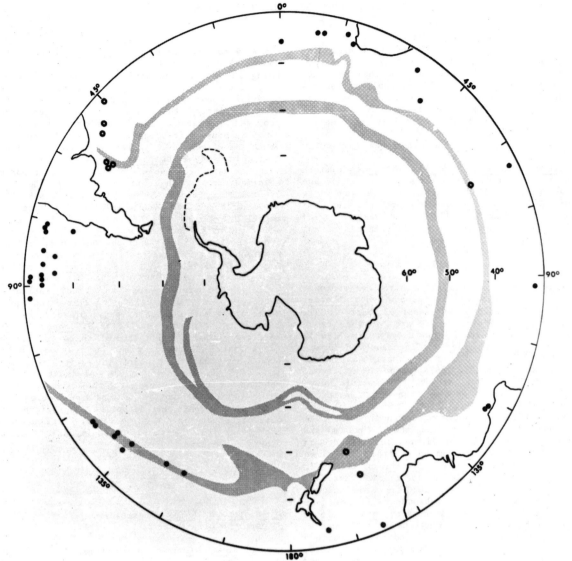

Fig. 22. Distributions of _Myctophum phengodes_ south of 30°S (dot) and of _Lampadena notialis_ (black dot with white center).

in the eastern tropical Pacific. _Eltanin_ specimens do not clearly fit the description of any of the species defined by Becker [1965]. In addition, there are morphological differences between specimens captured near Australia and those taken in the Pacific sector, and the material may include more than one species.

Myctophum Rafinesque, 1810

The genus _Myctophum_ is closely related to _Symbolophorus_ and includes approximately 15 species [Paxton, 1972; Moser and Ahlstrom, 1970]. Most species of _Myctophum_ are restricted to warm waters, although _M. punctatum_ is distributed in boreal waters of the North Atlantic Ocean, and one species is restricted to the study area. _Myctophum nitidulum_, which requires taxonomic clarification [Becker, 1967a; Nafpaktitis and Nafpaktitis, 1969], has been reported from just south of 30°S near 93°W [Craddock and Mead, 1970] but is not present in the study material.

Myctophum phengodes (Lütken, 1892)
Fig. 22

Myctophum phengodes Norman, 1930, p. 326 (D 87).--Nafpaktitis and Nafpaktitis, 1969, p. 26 (AB 7358).
Ctenoscopelus phengodes Becker, 1967b, p. 174 (near 33°S, 172°E).--Craddock and Mead, 1970, p. 26 (B 13; 16 stations 30°-35°S, 77°-95°W).

Size. _Myctophum phengodes_ attains a length of over 80 mm.
Distribution. Figure 22 shows the distribution of _M. phengodes_ south of 30°S. Additional records from 30°-35°S in the Atlantic and Indian sectors are reported in the literature [Lütken, 1892, Brauer, 1906; Barnard, 1925] but are not included in the figure. (Only Brauer's record from near 30°W would, if correct, significantly alter the apparent distribution.) _Myctophum phengodes_ is also known from as far north as 22°S in the western Indian Ocean [Nafpaktitis and Nafpaktitis,

1969] and as far north as 9°S in the eastern Indian Ocean [Legand and Rivaton, 1967]. An unver-
ified report of this species from the North Atlantic Ocean by Fowler [1901] probably represents
an error in labeling or identification. The
southern limits of M. phengodes generally correspond to the region of the subtropical convergence.
It may be absent in the western Atlantic sector.

Myctophum phengodes has often been taken in
surface waters.

Systematic notes. Myctophum phengodes differs
considerably from other members of the genus, and
Fraser-Brunner [1959] established a monotypic
genus, Ctenoscopelus, for this species. Nafpaktitis and Nafpaktitis [1969] and Paxton [1972]
indicate, however, that the differences do not
justify generic distinction.

Symbolophorus Bolin and Wisner

According to Bolin [1959], there are approximately 10 species in the genus Symbolophorus.
The closely related S. evermanni (Gilbert, 1905),
and S. rufinus (Tåning, 1928) are found in tropical waters [Nafpaktitis and Nafpaktitis, 1969].
Symbolophorus californiense (Eigenmann and Eigenmann, 1889) is restricted to transitional North
Pacific waters [Paxton, 1967a; Moser and Ahlstrom,
1970], and S. veranyi (Moreau, 1888) occurs north
of the Gulf Stream edge in the North Atlantic
Ocean [Backus et al., 1970]. Species in the
higher latitudes of the southern hemisphere have
been poorly understood, and their taxonomy is
extremely confused in the literature. Richardson
[1844] described Myctophum boops from a specimen
captured by the Erebus and Terror, he believed,
in the seas between Australia and New Zealand.
He also mentioned material of the same species
from the 'China Seas.' Subsequent authors considered M. boops to be a synonym of Gasteropelecus
humboldti Risso, 1810, which Esteve [1947] established and Bolin [1959] confirmed was actually a
synonym of Myctophum punctatum Rafinesque, 1810,
and reported specimens of what is now recognized
as the genus Symbolophorus from high southern
latitudes as Risso's species: Lütken [1892] from
Cape Horn; Gilbert [1911] from 39°S, 79°W; Waite
[1911] from New Zealand; Barnard [1925] from South
Africa; McCulloch [1929] from Lord Howe Island;
and Norman [1930] from the south Atlantic. Tåning
[1932], trusting Richardson's locality data,
recognized three allopatric subspecies on the
basis of differences in the structure of the
caudal luminous glands; M. humboldti humboldti
from the North Atlantic (the same as S. veranyi
(Moreau)); M. humboldti boops from the Tasman Sea;
and M. humboldti barnardi from near South Africa.
Tåning also commented on a discrepancy between his
observations on the structure of the supracaudal
glands of males from near Africa and those of the
specimens described by Barnard [1925], who Tåning
apparently believed had erred. Significantly,
Whitley [1953] discovered that the Erebus and Terror could not have been in the Tasman Sea when the
specimen described by Richardson as M. boops was
collected. He named a specimen from Lord Howe
Island Scopelus hookeri, mentioning the possibility that M. humboldti barnardi and M. humboldti
boops might be synonymous. Thus, in 1953, three

names were available for Symbolophorus in the
'Southern Ocean', the locality of the oldest name
being in doubt. The confused taxonomy, variability of caudal glands, and, primarily, inadequate
material apparently prevented Bolin and Wisner
from completing their study of the genus and
prompted Andriashev [1962] to place all southern
Symbolophorus, including his material from the
South Pacific, in the synonymy of the oldest
available name, S. boops (Richardson). Subsequent
authors have followed Andriashev and reported
specimens from south of 30°S as S. boops [Bussing,
1965; Becker, 1967a; Nafpaktitis and Nafpaktitis,
1969; Craddock and Mead, 1970]. The extensive
material examined in the course of the present
study indicates that six species of Symbolophorus
occur south of 30°S and that another, related
species exists in lower latitudes of the Atlantic
Ocean. The North Atlantic S. veranyi is related
to these species. Four of the six species, as
well as the species from the middle Atlantic,
require names. Ironically, however, inadequate
material from the South Atlantic prevents resolu-
tion of the taxonomic status of S. boops, S. barnardi, and S. hookeri, although there are indica-
tions that S. barnardi might be a synonym of S.
boops and that S. hookeri is valid. The most
important diagnostic feature of most of the
species is the structure of the caudal glands,
which in all but one or perhaps two species, are
sexually dimorphic and do not develop before members of either sex are quite large. This has made
identification of most juveniles impossible and of
others subjective.

Symbolophorus sp. A
Fig. 23

Symbolophorus boops Andriashev, 1962, p. 252
    (part, OB 415, female).

Size. Symbolophorus sp. A attains a length of
over 160 mm.

Distribution. Figure 23 shows the known distribution of S. sp. A. A number of juveniles
without developed caudal glands that probably
belong to this species are not included in the
figure, (Elt 165, 215, 1270, 1285, 1720; SOSC Elt
159, 92). Their localities would not alter the
apparent distribution of the species. Symbolophorus sp. A. has been collected between the Polar
Front and the subtropical convergence in the Pacific sector within the counterclockwise flow of
subantarctic waters. Its range is bordered by
the ranges of S. sp. B. and S. sp. D to the east
and north, respectively.

Symbolophorus sp. A has been taken in the surface.

Systematic notes. Characters which distinguish
S. sp. A from its congeners are a low number of
rakers on the first gill arch (18-20), a high
number of Ao photophores (17-18) and anal - fin
rays (22-23), and the unique morphology of the
caudal luminous glands. The male supracaudal
gland covers most of the dorsal surface of the
caudel peduncle, and the female has at least four
luminous plates on the ventral surface of the
caudal peduncle.

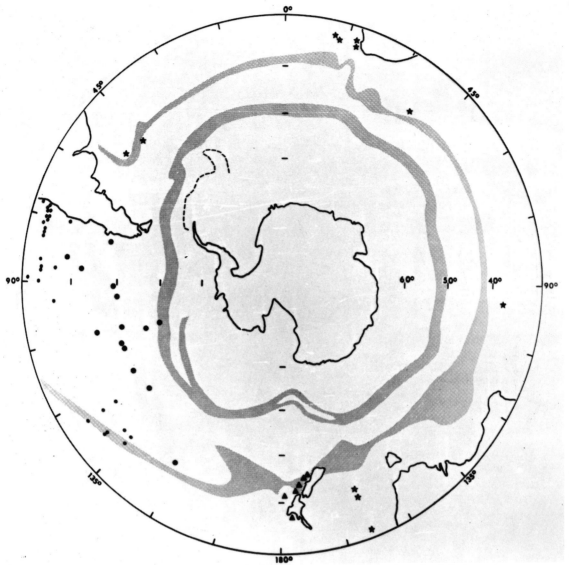

Fig. 23. Distributions of <u>Symbolophorus</u> sp. A (large dot) and <u>Symbolo-</u>
<u>phorus</u> sp. C (triangle) and of <u>Symbolophorus</u> sp. D (small dot), and <u>Sym-</u>
<u>bolophorus</u> spp. (star) south of 30°S.

<u>Symbolophorus</u> sp. B
Fig. 24

<u>Myctophum humboldti</u> Norman, 1930, p. 325 (part, D 7).
<u>Symbolophorus boops</u> Craddock and Mead, 1970, p. 30 (part, B 13; see Appendix 1).

<u>Size.</u> <u>Symbolophorus</u> sp. B attains a length of over 135 mm.
<u>Distribution.</u> Figure 24 shows the known distribution of <u>S</u>. sp. B south of 30°S. An additional specimen has been collected in the Benguela Current near 25°S, 13°E (WS 1070). <u>Symbolophorus</u> sp. B is distributed in the region of the subtropical convergence from New Zealand west to the Atlantic sector, in the Benguela Current, and between 35°S and 50°S off Chile. It is apparently absent in the central-eastern Pacific sector.

<u>Symbolophorus</u> sp. B has often been taken in the surface.
<u>Systematic notes.</u> Characters which distinguish <u>S</u>. sp. B from its congeners are a high number of rakers on the first gill arch (more than 20), a low number of Ao photophores (14-15) and anal – fin rays (20-22), the length of the upper jaw relative to the distance from the last dorsal fin ray to the adipose fin (less than one to one) and the diameter of the orbit (greater than two to one), the presence of darkly pigmented areas on the pectoral fins of specimens larger than 65 mm, and the morphology of the caudal luminous glands. The adult male supracaudal gland consists of three coalesced plates which do not cover more than one – half of the dorsal surface of the caudal peduncle. Adult females characteristically have two or three infracaudal plates. Individuals captured off Chile are more slender than those from

34

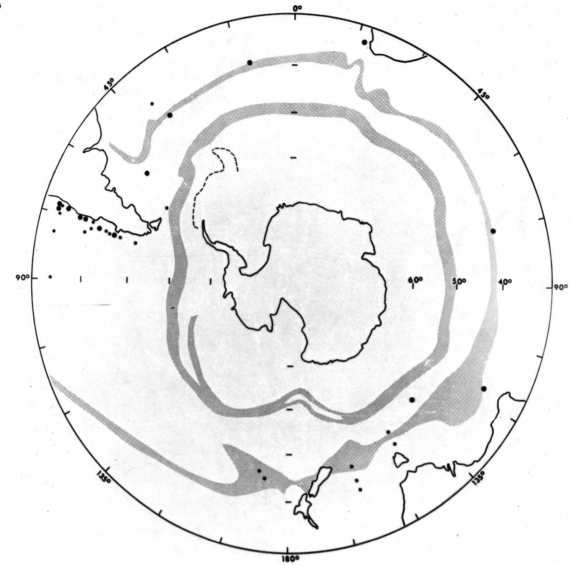

Fig. 24. Distribution of Symbolophorus sp. B south of 30°S. Small dot indicates individuals without developed caudal luminous glands and large dot indicates individuals with developed caudal luminous glands.

the western population. Symbolophorus sp. B. is most similar to S. sp. C.

Symbolophorus sp. C
Fig. 23

Myctophum humboldti Waite, 1911, p. 166 (near New Zealand).

Size. Symbolophorus sp. C attains a length of at least 90 mm.

Distribution. Figure 23 shows the known distribution of S. sp. C. This species has been collected only near the east coast of New Zealand and often on the surface.

Systematic notes. Characters which distinguish S. sp. C. from its congeners include a high number of rakers on the first gill arch (14-15), the length of the upper jaw relative to the distance from the last dorsal fin ray to the adipose fin (about one to one) and the diameter of the orbit (about two to one), the absence of darkly pig-

mented areas on the pectoral fins, and the morphology of the caudal luminous glands. The caudal glands are similar to those of S. sp. B.

Symbolophorus sp. D
Fig. 23

Symbolophorus boops Andriashev, 1962, p. 252 (part, OB 442).--Craddock and Mead, 1970, p. 30 (part, B 13; see Appendix 1).

Size. Symbolophorus sp. D attains a length of at least 110 mm.

Distribution. The known distribution of Symbolophorus sp. D south of 30°S is shown in Figure 23. Additional specimens collected by Atlantis cruise 221 near 25°S, 71°W are present in the fish collection of the Scripps Institution of Oceanography. This species is distributed in the region of the subtropical convergence only in the Pacific sector east of 135°W, and has frequently been taken in the surface.

Systematic notes. Characters which distinguish
S. sp. D from its congeners include a low number
of rakers on the first gill arch (17-20), a high
number of AO photophores (usually 17-20), and the
morphology of the caudal luminous glands. Males
and females characteristically have two luminous
plates on both the dorsal and ventral surfaces of
the caudal peduncle. This species appears to be
most similar to S. hookeri.

Symbolophorus spp.

Myctophum humboldti Norman, 1930, p. 325 (part, D
71).
Myctophum humboldti barnardi Tåning, 1932, p. 128
(near South Africa).
Scopelus hookeri Whitley, 1953, p. 134 (Lord Howe
Island).

Distribution. Figure 23 shows the distribution
of forms assigned to S. spp. They are distributed
in and north of the region of the subtropical
convergence except in the Pacific sector east of
New Zealand, where they are absent.
Systematic notes. Symbolophorus spp. includes
S. boops (Richardson), S. barnardi (Tåning), and
S. hookeri (Whitley), none of which, as was mentioned earlier, can be properly defined with
available material. The type of S. boops is male
with a single supracaudal gland, and the type of
S. barnardi is a female with two overlapping
supracaudal and three overlapping infracaudal
plates. Tåning [1932] remarked that the males he
examined from near South Africa had only supracaudal glands and that the females had both infracaudal and supracaudal glands. He disagreed
with Barnard [1925], who had reported both males
and females with dorsal and ventral luminous
glands on the caudal peduncle. I suspect that
had Tåning known that the holotype of S. boops
could not have come from the Tasman Sea, as was
reported by Richardson [Whitley, 1953], but probably came from the South Atlantic near South Africa [Andriashev, 1962], he would have refrained
from introducing a new name or possibly retained
Richardson's name for the African population.
Tåning [1932] further described males and females
from New Zealand waters (his M. humboldti boops)
as having only supracaudal and infracaudal plates,
respectively. It seems almost certain that the
specimens on which he based these observations
were actually S. sp. C. Symbolophorus sp. C were
collected on the Dana in 1929 (station 3644), and
these specimens were probably available to Tåning.
His description could also have been based on the
type description of S. boops, the S. sp. C reported as M. humboldti by Waite [1911], or on
additional material. Thus, while Whitley [1953,
p. 135] was correct in pointing out that the type
of S. boops could not have come from the Tasman
Sea, he was incorrect in stating that S. hookeri
with 'one to three small luminous scales above or
below caudal peduncle' was the same as the M. humboldti boops of Tåning: the holotype of S. hookeri
has one supracaudal gland (much smaller than that
of S. sp. C or S. boops) and a damaged infracaudal
region. The paratypes (one male and two females),
also from Lord Howe Island, have 1-2 supracaudal
and 2-3 infracaudal plates (R. L. Bolin and R. L.
Wisner, unpublished manuscript, 1971), unlike any
material then known from the Tasman Sea but

similar to the males and females of M. humboldti
reported by Barnard from South Africa. Additional
specimens of both sexes from Lord Howe Island (BM
1926.6.30.6-8), from near 40°S in the Tasman Sea
(Elt 1822, 2223), and from South Africa (D 1768;
BM 1922.1.13.61) all have the hookeri-type caudal
glands. A male and female from a collection near
South Africa have caudal glands similar to those
of S. boops and S. barnardi, respectively (SAM:
west of Cape Point). Considerable variability in
caudal gland morphology of specimens from other
localities obscure the picture. However, most of
these specimens are relatively small and their
caudal glands are probably not fully developed.
Several females less than 70 mm long (including
one gravid specimen) that have single dorsal and
ventral luminous plates are exceptional and
represent a distinct population that occurs in
the lower latitudes of the Atlantic Ocean (WS
1100).
The preceding observations prompt several suggestions. The general similarity of the holotype
of S. boops and S. boops like specimens to the
holotype of S. barnardi and S. barnardi like
specimens (except for sex and caudal gland morphology), Taning's [1932] belief that similar
forms made up a South African species, the occur-
ence of both forms in a single sample (SAM: west
of Cape Point), and the high probability that
both holotypes were collected in the same area
lead me to suspect that S. boops and S. barnardi
are synonymous. The consistently distinct caudal
gland pattern of large specimens captured near
South Africa indicates that S. hookeri is valid
and that it extends from the Tasman Sea to Africa.
The consistently distinct caudal gland morphology
of relatively small specimens, including a gravid
female 68 mm long, captured in low latitudes of
the Atlantic indicated the existence there of yet
another species. Symbolophorus boops and its
relatives pose a difficult taxonomic problem which
can only be resolved by analysis of more extensive
material than was available for the present study.

Gonichthys Gistel, 1850

Becker [1964a] recognizes four species of Gonichthys: G. coccoi (Cocco), in the tropical and
subtropical Atlantic Ocean; G. tenuiculus (Garman), in the eastern tropical Pacific Ocean; G.
venetus Becker, in the tropical Pacific Ocean;
and G. barnesi Whitley, in higher middle latitudes of the southern hemisphere. Close similarity of these forms to each other and considerable
intraspecific variation make their identification
very difficult. Gonichthys barnesi is accepted
as the name for the form present in the study
material.

Gonichthys barnesi Whitley, 1943
Fig. 25

Myctophum coccoi Norman, 1930, p. 325 (D 87, 247).
Gonichthys barnesi Whitley, 1943, p. 174 (Lord
Howe Island).--Becker, 1964a, p. 35 (17 localities in the Pacific and Indian oceans.--Craddock
and Mead, 1970, p. 27 (B 13, 12 stations 30°-
35°S, 77°-95°W).

Size. Gonichthys barnesi grows to a length of
over 50 mm.

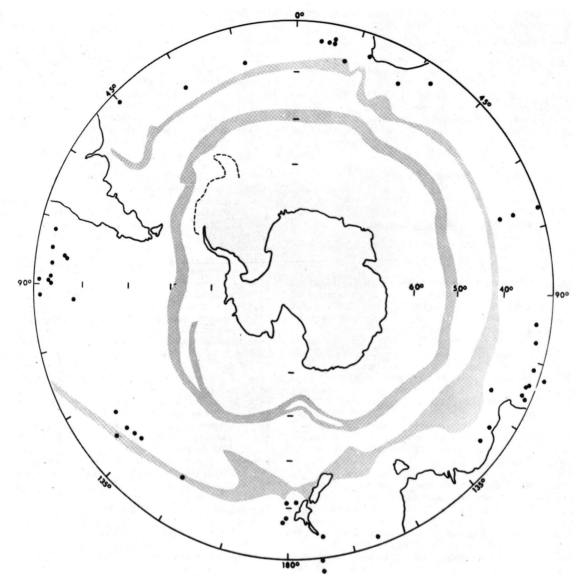

Fig. 25. Distribution of Gonichthys barnesi.

Distribution. Figure 25 shows the distribution of G. barnesi. This species is distributed in the region of the subtropical convergence and extends 12°-15° farther north into subtropical waters. Additional records from the literature [see Becker, 1964a] that are not included in the figure do not alter its apparent distribution.

Gonichthys barnesi has frequently been taken on the surface.

Loweina Fowler, 1925

The genus Loweina requires revision. Four very similar species are currently recognized: L. rara (Lütken), known from between 23°S and 36°N in the Atlantic Ocean [Becker, 1964a; Wisner, 1971]; L. terminata Becker, found in the subtropical north Pacific [Becker, 1964a]; L. laurae Wisner, which occurs between 30°N and 30°S in the eastern Pacific Ocean; and L. interrupta (Tåning), recorded from the Atlantic and south Indian oceans [Nafpaktitis and Nafpaktitis, 1969]. The status of the type species is ambiguous. Lütken [1892] apparently illustrated a species other than the

species he described as Scopelus rarus. The figured species in Lütken's description was later named M. interruptum in a key to myctophids of the North Atlantic [Tåning, 1928]. Tåning's interpretation has been accepted by recent workers.

Loweina interrupta (Tåning, 1928)
Fig. 26

Myctophum interruptum Norman, 1930 (D 87, 257).
Loweina interrupta Nafpaktitis and Nafpaktitis, 1969, p. 31 (AB 7127).

Figure 26 shows the known distribution of L. interrupta south of 30°S. This species is also known from middle latitudes of the North Atlantic [Nafpaktitis and Nafpaktitis, 1969]. Material reported from the equatorial Atlantic by Brauer [1906] and considered by Tåning [1928] to be L. interrupta should be reexamined. In the southern hemisphere, L. interrupta has been collected near the northern limits of the subtropical convergence.

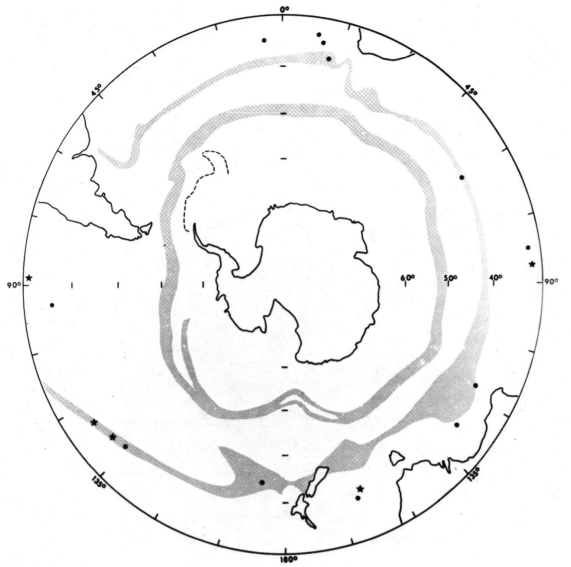

Fig. 26. Distribution of <u>Loweina</u> <u>interrupta</u> (star) and <u>Notolychnus</u> <u>vald-</u> <u>iviae</u> (dot) south of 30°S.

<u>Notolychnus</u> Fraser-Brunner, 1949

The monotypic genus <u>Notolychnus</u> is placed in its own tribe by Paxton [1972].

<u>Notolychnus</u> <u>valdiviae</u> (Brauer, 1904)
Fig. 26

<u>Myctophum</u> <u>valdiviae</u> Brauer, 1906, p. 206 (near 32°S, 85°E).
<u>Notolychnus</u> <u>valdiviae</u> Craddock and Mead, 1970, p. 30 (B 13-29).

<u>Size</u>. <u>Notolychnus</u> <u>valdiviae</u> is a diminutive species which reaches a length of about 23 mm [Bolin, 1959].
<u>Distribution</u>. Figure 26 shows the known dis-tribution of <u>N</u>. <u>valdiviae</u> south of 30°S. This species is also known from the western Atlantic Ocean between 30°S and 35°N and from the eastern Atlantic between 50°N and 5°S at about 20°W [Bra-uer, 1906; Bolin, 1959; Becker, 1967a; Backus et al., 1970; <u>Discovery</u> data]; from one locality near 27°S, 6°E in the southeastern Atlantic [Bra-uer, 1906]; from the Indian Ocean between 30°S and 12°N [Brauer, 1906; Legand, 1967; Nafpaktitis and Nafpaktitis, 1969]; from the equatorial Pacific Ocean [Grandperrin and Rivaton, 1966]; and from the eastern Pacific Ocean between 30°S and 32°N [Berry and Perkins, 1966; Alhstrom, 1971; Craddock and Mead, 1970]. <u>Notolychnus</u> <u>valdiviae</u> is generally distributed throughout the warm waters of the World Ocean. The northern boundar-ies of the region of the subtropical convergence approximate southern limits of distribution.

<u>Notolychnus</u> <u>valdiviae</u> has been collected often at depths of less than 100 m.

<u>Lampadena</u> Good and Bean, 1896

Krefft [1970] discerns three phylogenetic trends within the genus <u>Lampadena</u>. The <u>L</u>. <u>luminosa</u> species-group includes <u>L</u>. <u>luminosa</u>, found in equa-torial and adjacent waters of all oceans, <u>L</u>. <u>anomala</u>, known from equatorial waters of the Atlantic Ocean, and <u>L</u>. <u>urophaos</u>, which occurs

38

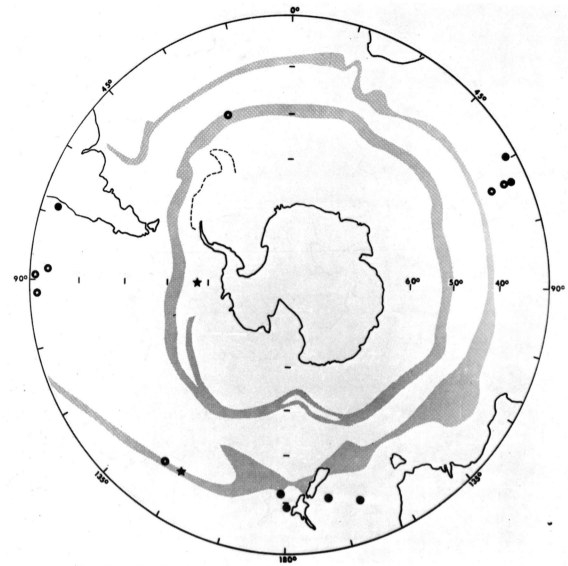

Fig. 27. Distributions of Lampadena speculigera (dot) and of Lampadena dea (black dot with white center) and Taaningichthys bathyphilus (star) south of 30°S.

between 20°N and 37°N in the northwestern Atlantic and from 25°N to 41°N in the northeastern Pacific Ocean [Nafpaktitis and Paxton, 1968; Krefft, 1970]. The L. speculigera species-group includes the three species discussed below and L. pontifex, a species known only from the Atlantic near Africa at about 15°N [Krefft, 1970]. Lampadena chavesi is distinct from the other members of the genus and is known from between 30°N and 45°N in the North Atlantic and from 25°S to 35°S in the southwestern Atlantic, western Indian, and southeastern Pacific oceans [Krefft, 1970; Nafpaktitis and Paxton, 1968].

Lampadena speculigera Goode and Bean, 1896
Fig. 27

Lampadena speculigera Nafpaktitis and Paxton, 1968, p. 10 (Elt 1402, 1710, 1820, AB 157, 352b; B 13, 54).

Figure 27 shows the known distribution of L. speculigera south of 30°S. Although Bolin [1959]

states that L. speculigera is distributed between 13°N and 51°N across the north Atlantic, a plot of the specimens included in his synonymy and others reported subsequently [Becker, 1967a; Nafpaktitis and Paxton, 1968] indicates that it occurs between 35°N and 45°N west of 30°W. Bolin also includes the L. braueri reported by Norman [1930] from near 16°S, 11°E in the synonymy of L. speculigera. If he is correct, and Norman's figure and description indicate that he is, this would be the only known record of the species from the South Atlantic. It is apparently absent in the southwestern Atlantic [Krefft, 1970].

Lampadena notialis Nafpaktitis and Paxton, 1968
Fig. 22

Lampadena notialis Nafpaktitis and Paxton, 1968, p. 13 (AB 7339; ELT 1830, 1841).--Krefft, 1970, p. 277 (six stations in the southwestern Atlantic Ocean).

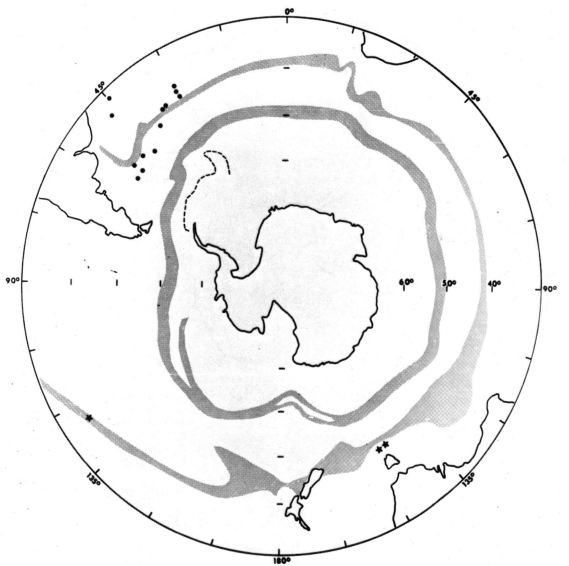

Fig. 28. Distribution of <u>Lepidophanes</u> <u>guentheri</u> south of 30°S (dot) and of <u>Bolinichthys</u> <u>supralteralis</u> in the southern hemisphere (star).

Figure 22 shows the known distribution of <u>L</u>. <u>notialis</u>. This species is distributed in the region of the subtropical convergence in the Indian, Tasman, and southwestern Atlantic sectors of the Southern Ocean. It appears to be absent from the Pacific sector.

<u>Lampadena dea</u> Fraser-Brunner, 1949
Fig. 27

<u>Lampadena</u> <u>dea</u> Fraser-Brunner, 1949, p. 1101 (D 395).--Nafpaktitis and Paxton, 1968, p. 36 (AB 7085, 7324; B 13 22, 28, 30).

Figure 27 shows the distribution of <u>L</u>. <u>dea</u> south of 30°S. <u>Lampadena dea</u> is also known from near 25°S in the southeastern Atlantic Ocean. This species has been collected in and north of the region of the subtropical convergence in the Pacific, Indian, and southeastern Atlantic sectors of the Southern Ocean. It is apparently absent in the southwestern Atlantic sector [Krefft, 1970].

<u>Taaningichthys</u> Bolin, 1959

The genus <u>Taaningichthys</u> includes three species broadly distributed in warm waters of the World Ocean [Davy, 1972]. One species, <u>T</u>. <u>bathyphilus</u>, is present in the study collection.

<u>Taaningichthys bathyphilus</u> (Tåning, 1928)
Fig. 27

The two known records of <u>T</u>. <u>bathyphilus</u> from south of 30°S are shown in Figure 27. The unusual occurrence of this species south of the Polar Front near 90°W may be due to southward transport in deeper layers of water. This species has only been collected in trawls which have fished deeper than about 575 m [Davy, 1972]. <u>Taaningichthys bathyphilus</u> is also known from between 29°S and 10°N in the Indian Ocean, from the North Atlantic between 10°N and 40°N, and from as far north as 40°N in the Pacific Ocean [Davy, 1972].

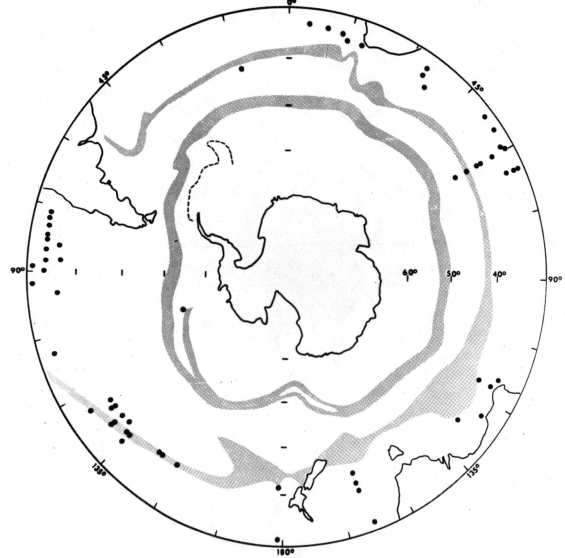

Fig. 29. Distribution of _Ceratoscopelus warmingi_ south of 30°S.

Bolinichthys Paxton, 1972

The genus Bolinichthys is related to <u>Lepidophanes</u> and <u>Cerstoscopelus</u> [Paxton, 1972]. Most of the more than five species included in the genus are restricted to warm waters at low latitudes. <u>Bolinichthys</u> <u>indicus</u>, described from material collected between 22°S and 45°S in the Indian Ocean [Nafpaktitis and Nafpaktitis, 1969] and <u>B. supralateralis</u>, present in the study material, are exceptions.

Bolinichthys supralateralis (Parr, 1928)
Fig. 28

The three known records of <u>B. supralateralis</u> from the southern hemisphere are shown in Figure 28. It had previously been reported only from the type locality near the Bahama Islands in the North Atlantic. Backus et al. [1970] suggest that it may have a tropical distribution pattern in the North Atlantic Ocean.

Lepidophanes Fraser-Brunner, 1949

Paxton [1972] recognizes two species of <u>Lepidophanes</u> and indicates that this genus is most closely related to <u>Ceratoscopelus</u>. <u>Lepidophanes gaussi</u> (Brauer) is distributed in the Atlantic Ocean between 30°N and 23°S [Bolin, 1959; Becker, 1967a]. The other species, <u>L. guentheri</u>, is present in the study collections.

Lepidophanes guentheri (Goode and Bean, 1896)
Fig. 28

<u>Lampanyctus guentheri</u> Norman, 1930, p. 329 (D 66, 69, 76, 240, 241, 242).

<u>Size</u>. <u>L. guentheri</u> attains a length of at least 65 mm.

<u>Distribution</u>. Figure 28 shows the known distribution of <u>L. guentheri</u> south of 30°S. This species is also known from between 30°S and 38°N

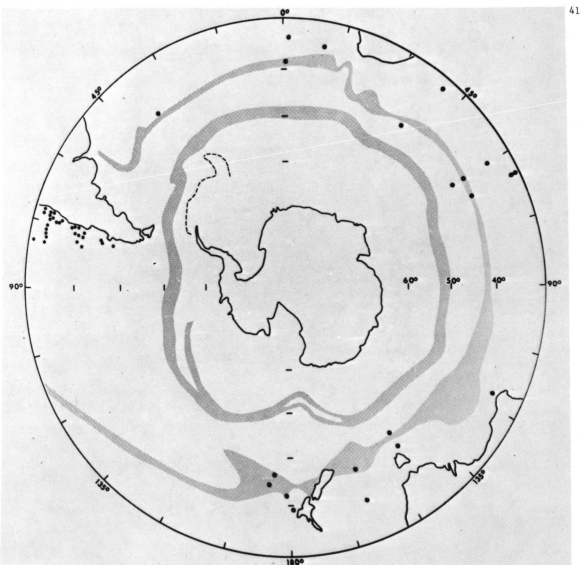

Fig. 30. Distribution of Lampanyctus niger-ater complex south of 30°S
(large dot) and of Lampanyctus iselinoides (small dot).

in the western Atlantic and extends eastward
across the Atlantic Ocean between 18°N and 18°S
[Bolin, 1959; Backus et al., 1969; Becker, 1967;
Discovery data].

Lepidophanes guentheri has often been collected at depths less than 100 m.

Ceratoscopelus Günther, 1864

...scopelus is closely related to
...1972]. Ceratoscopelus
...ecies of the genus
...e Atlantic Ocean
...[Backus et al.,
...paktitis [1969]
...y related species:
...orth Pacific Ocean,
...ntic, Indian, and,
...s. Ahlstrom [1971]
...rvae from the Indian
...he distinction of the

two species as defined by Nafpaktitis and Nafpaktitis [1969].

Ceratoscopelus warmingi (Lütken, 1892)
Fig. 29

Lampanyctus townsendi McCulloch, 1923, p. 115
(Lord Howe and Kermadec Islands).--Norman, 1930,
p.327 (part, D 104, 270, 284, 285, 286, 287,
294, 296, 297).
Ceratoscopelus townsendi Andriashev, 1962, p. 260
(OB 409, 424).--Craddock and Mead, 1970, p. 26
(B 13, 16 stations 77°-93°W, 30°-34°S).
Ceratoscopelus warmingi Nafpaktitis and Nafpak-
titis, 1969, p. 63 (AB 14 stations 30°-45°S,
60°-65°E).

Size. Ceratoscopelus warmingi attains a length
of at least 70 mm.
Distribution. Figure 29 shows the distribution
of C. warmingi south of 30°S. Additional locali-

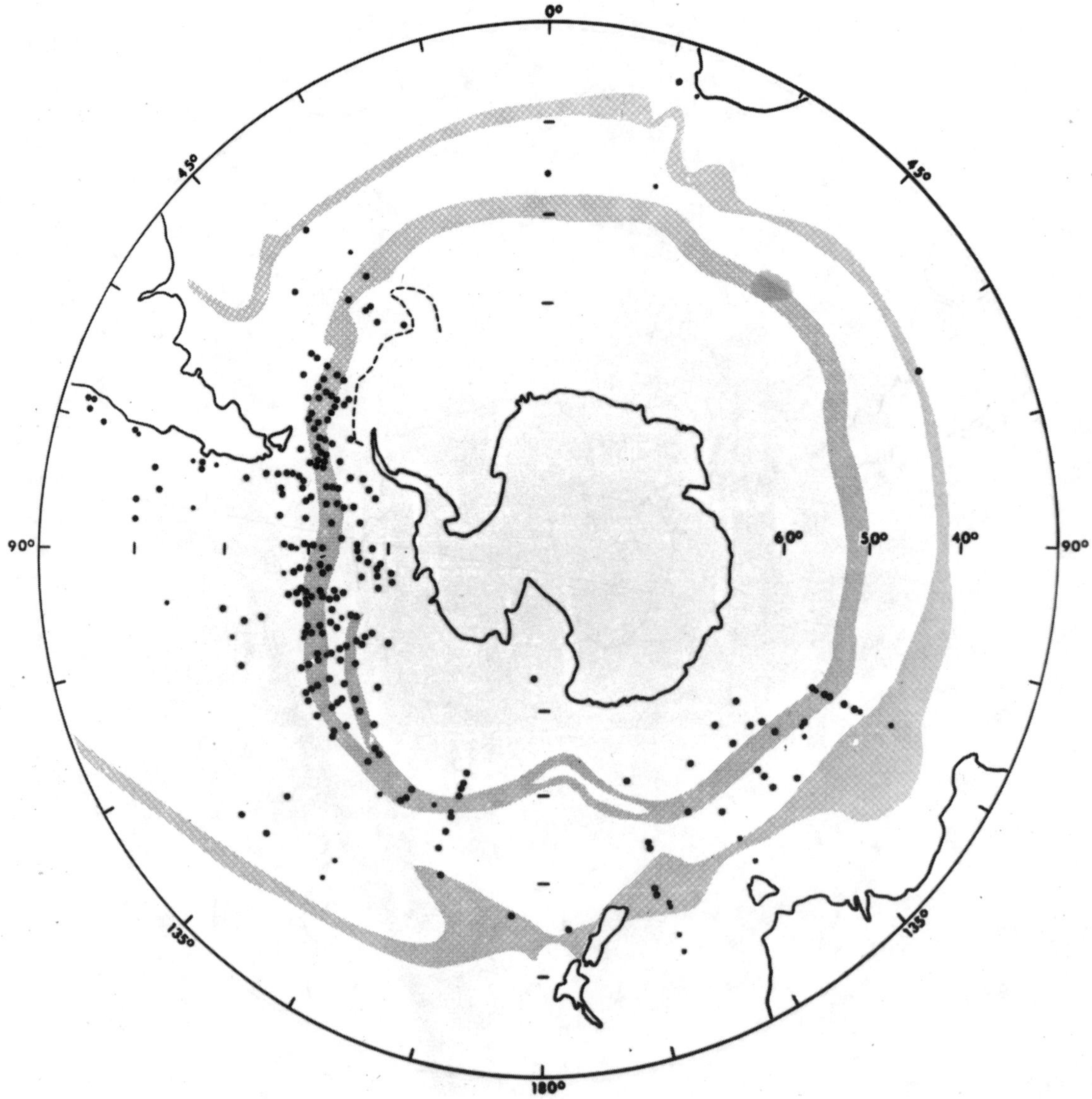

Fig. 31. Distribution of Lampanyctus achirus south of 30°S. Individuals smaller than 60 mm long are indicated by small dot and individuals larger than 59 mm long are indicated by large dot.

ties reported from south of 30°S [Pappenheim, 1914; Barnard, 1925; Norman, 1930, part D 80, 85, 257] that are not included in the figure would not alter the apparent distribution. Ceratoscopelus warmingi is also known from between 30°S and 40°N in the Atlantic Ocean [Bolin, 1959; Backus et al., 1965, 1969, 1970; Becker, 1967a; Discovery data] and from between 30°S and 12°N in the Indian Ocean [Nafpaktitis and Nafpaktitis, 1969]. Larvae which may be conspecific with C. warmingi have been reported from north of 20°S in the eastern Pacific Ocean [Ahlstrom, 1971]. Krefft [1974] reports this species from a broad area of the South Atlantic Ocean north of 40°S. The anomalous record from near the Polar Front at about 105°S is that of a specimen reported by Andriashev [1962]. This fish has often been collected at depths less than 100m.

Lampanyctus Bonaparte, 1840

The relationships of approximately 40 species of Lampanyctus are not clear [Paxton, 1972]. At least 13 species of this genus are present in the study collections. The remaining species are mostly restricted to warm waters. Lampanyctus parvicauda, which is endemic to eastern Pacific waters, has been reported from just south of 30° near the Chilean coast [Bussing, 1965].

Lampanyctus niger-ater complex
Fig. 30

Lampanyctus niger Norman, 1930, p. 331 (part, 81, 241).

Lampanyctus ater Nafpaktitis and Nafpak 1969, p.44 (AB 7105, 7110, 7133, 7351 7357).

Distribution. The distribution of t complex south of 30°S is shown in Fi but one of the records of L. at Norman [1939] from the South Atl mined and found to be L. achirus below). Some of the L. niger

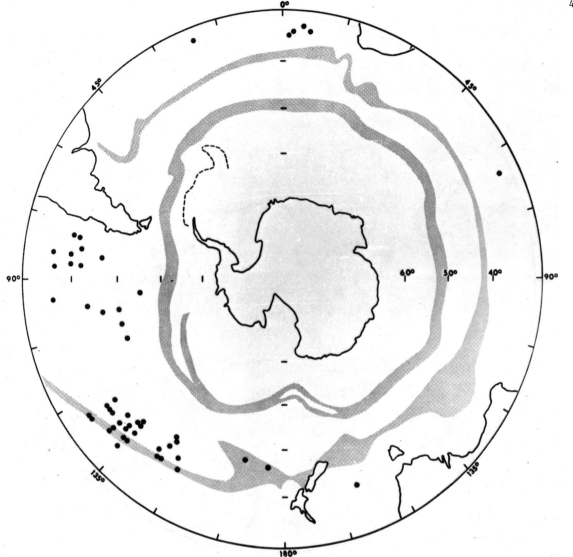

Fig. 32. Distribution of *Lampanyctus* sp A.

be L. sp. A. Additional reports of L. niger by Brauer [1906], Pappenheim [1914], and Barnard [1925] from south of 30°S are not included in the figure because of the possibility that L. achirus or L. sp. A might be included in their material. These localities would not, however, significantly alter the pattern evident in Figure 30. The complex is not known from east of 165°W in the southern hemishere. West of 165°W the southern limits of distribution correspond to the region of the subtropical convergence.

Systematic notes. *Lampanyctus niger* was described from the Celebes Sea by Günther [1887], and the name has since been applied to populations in the Atlantic Ocean [Brauer, 1906; Pappenheim, 1914; Barnard, 1925; Norman, 1930], the North Pacific Ocean [Gilbert, 1905; Kulikova, 1960; Berry and Perkins, 1966], and the Indian Ocean [Brauer, 1906]. *Lampanyctus ater* was described from the North Atlantic Ocean by Tåning [1969]. Bolin [1959], while questioning the validity of L. ater, retained the name for the North Atlantic population although he did not include material reported by Pappenheim [1914] and Brauer [1906] from the equatorial and North Atlantic in his

synonymy. He indicated that Indian Ocean and southern populations are characterized by a reduced number of plates in their caudal luminous glands in comparison to the plates of L. ater. Nafpaktitis and Nafpaktitis [1969] reported L. ater and two related species from the Indian Ocean. Their L. sp. 1 is similar to L. niger, characterized by Bolin [1959]. Eltanin and Discovery material is similar to their L. ater.

Lampanyctus achirus Andriashev, 1962
Fig. 31

Lampanyctus ater Norman, 1930, p. 331 (part, D101, 239).
Lampanyctus achirus Andriashev, 1962, p. 256 (OB 401, 409, 455, 464).--Bussing, 1965, p. 203 (Elt 190, 742, 743).--Becker, 1967a, p. 116 (near 42°S, 39°W).--Nafpaktitis and Nafpaktitis, 1969, p. 45 (part, AB 7339).

Size. *Lampanyctus achirus* attains a length of over 150 mm.
Distribution. The distribution of L. achirus south of 30°S is shown in Figure 31. The speci-

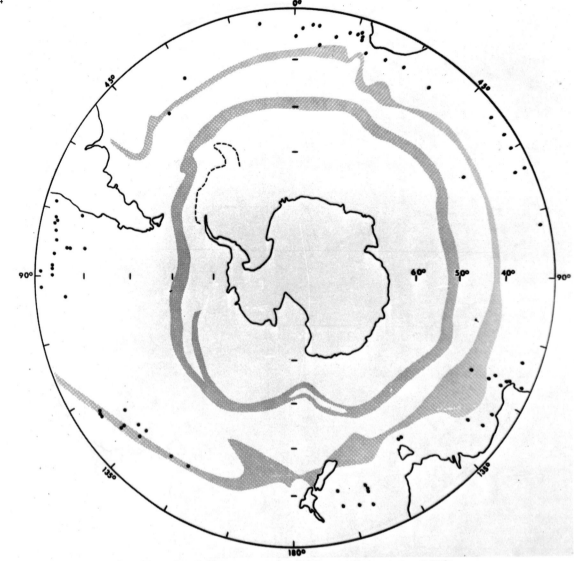

Fig. 33. Distribution of _Lampanyctus pusillus_ south of 30°S.

mens reported by Craddock and Mead [1970] as L. achirus are not included in the figure because their material almost certainly included the smaller L. sp. A, which apparently does not grow larger than 90 mm long. Craddock and Mead report specimens larger than 90 mm only east of 78°W. These specimens undoubtedly are L. achirus. Lampanyctus achirus is known from north of 30° only near South America, where it has been reported from as far north as 8°S off the Peruvian coast [Bussing, 1965].

The northern limits of distribution of L. achirus correspond to the region of the subtropical convergence and the southern distributional limits of L. sp. A, except in the eastern Pacific where it is apparently carried northward in deeper layers of the Peru Current. Individuals less than 60 mm long have been collected as far south as the Polar Front, whereas larger individuals have been taken farther south, as far as the Weddell-Scotia confluence and the northern reaches of the Ross Sea (Elt 2111). The latter record is the only specimen captured in numerous trawls taken in the north of the Ross Sea.

Lampanyctus achirus is a deep-living species. It was captured many times by the _Eltanin_, but only four times in trawls which did not fish deeper than 400 m.

Systematic notes. _Lampanyctus achirus_ may be distinguished from the related L. sp. A by the characters listed under the latter species.

Lampanyctus sp.A
Fig. 32

Lampanyctus intricarius Norman, 1930, p. 330 (part, D 87).
Lampanyctus ater Norman, 1930, p. 331 (part, D 86).
Lampanyctus niger Norman, 1930, p. 331 (part, D 81, 252).
Lampanyctus achirus Nafpaktitis and Nafpaktitis, 1969, p. 45 (part, AB 7339).

Size. Lampanyctus sp. A attains a length of about 85 mm. Gravid females as small as 56 mm were collected from June to September.
Distribution. Figure 32 shows the distribution

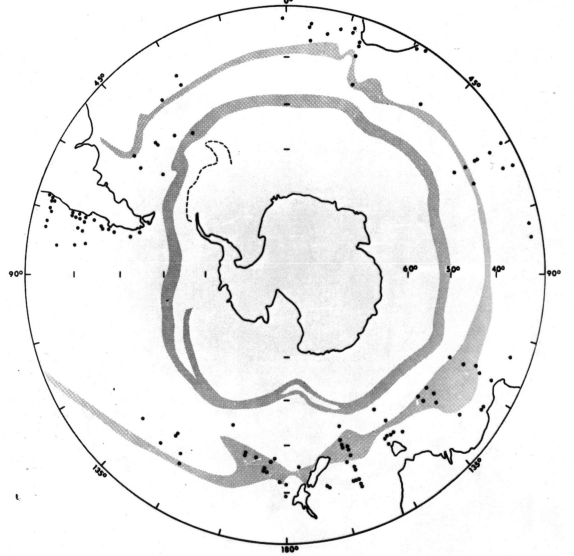

Fig. 34. Distribution of <u>Lampanyctus</u> <u>australis</u> south of 30°S.

of <u>L</u>. sp. A. As was discussed above, the material reported as <u>L</u>. <u>achirus</u> by Craddock and Mead [1970] from the southeastern Pacific almost certainly includes <u>L</u>. sp. A. This species has been collected in and north of the region of the subtropical convergence in the Atlantic, Indian, and Tasman sectors and in and south of the subtropical convergence in the Pacific sector. Its southern limits of distribution correspond to the northern limits of <u>L</u>. <u>achirus</u>.

<u>Systematic notes</u>. <u>L</u>. sp. A may be distinguished from the closely related <u>L</u>. <u>achirus</u> by a lower number of rakers on the upper limb of the first gill arch (4 versus 5), a greater proportional distance between the snout and the origin of the dorsal fin (more than versus less than one-half the standard length), and its smaller size at maturity.

<u>Lampanyctus pusillus</u> (Johnson, 1890)
Fig. 33

<u>Lampanyctus</u> <u>pusillus</u> Norman, 1930, p. 330 (part, D 85, 86, 89, 100, 101, 257).--Bussing, 1965, p. 207, (Elt 80).--Becker, 1967a, p. 109 (near 42°S, 39°W).--Nafpaktitis and Nafpaktitis,

1969, p. 52 (AB 7138, 7324, 7354, 7357).--Craddock and Mead, 1970, p. 29 (B13, 16 stations: 77°-94°W, 30°-34°S).

<u>Size</u>. Lampanyctus pusillus attains a length of at least 34 mm. Gravid females about 32 mm long were collected in July, August, and December.

<u>Distribution</u>. Figure 33 shows the distribution of <u>L</u>. <u>pusillus</u> south of 30°S. This species is also known from between 20°N and 60°N in the North Atlantic Ocean [Bolin, 1959; Becker, 1967a; Backus et al., 1969, 1970], from the Indian Ocean as far north as 25°S [Nafpaktitis and Nafpaktitis, 1969], and from one locality in the South Atlantic Ocean near 18°S, 4°E (D 1602). The study area includes the southern limit of distribution of <u>L</u>. <u>pusillus</u>. The southernmost records are in or very close to the subtropical convergence.

<u>Lampanyctus</u> <u>pusillus</u> has often been taken at depths less than 100 m.

<u>Lampanyctus australis</u> Tåning, 1932
Fig. 34

<u>Lampanyctus</u> <u>alatus</u> Norman, 1930, p. 330 (part, D 101, 239, 257); 1957, p. 86 (BANZARE 111).

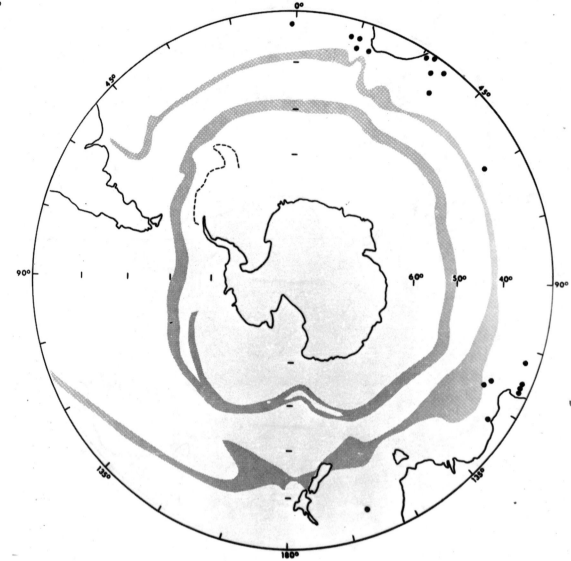

Fig. 35. Distribution of Lampanyctus alatus south of 30°S.

Lampanyctus australis Bussing, 1965, p. 203 (Elt 742).--Becker, 1967a, p.11 (near 42°S, 39°W).--Nafpaktitis and Nafpaktitis, 1969, p. 54 (AB 7100, 7107, 7118, 7121, 7127, 7133, 7138, 7140, 7324, 7351, 7354).--Craddock and Mead, 1970, p. 28 (B13 4, 5, 6, 10, 16, 40, 41, 43, 46, 47, 49, 53, 54, 58).

Size. Lampanyctus australis attains a length of at least 110 mm. Gravid females about 95 mm long were collected in June, September, and October.

Distribution. Figure 34 illustrates the known distribution of L. australis south of 30°S. This species occurs in the region of the subtropical convergence, except in the central Pacific sector where it is absent, and off Chile where it has been collected between 35°S and 55°S. Lampanyctus australis has also been taken north of 30°S in the Benguela Current near 18°S, 4°E (D 1602) and in the Agulhas Current at about 27°S, 39°E (D 1571).

Lampanyctus australis has often been taken at depths less than 100 m.

Lampanyctus alatus Goode and Bean, 1896
Fig. 35

Lampanyctus alatus Norman, 1930, p. 330 (part, D 281, 284, 285, 286, 294, 296, 297).--Nafpaktitis and Nafpaktitis, 1969, p. 53 (AB 7121).

Size. Lampanyctus alatus, in contrast with the related L. australis, attains a length of only about 55 mm. Nafpaktitis and Nafpaktitis [1969] have found gravid females about 40 mm long.

Distribution. The distribution of L. alatus south of 30°S is shown in Figure 35. Additional reports of this species from south of 30°S [Brauer, 1906; Barnard, 1925] are not included in the figure because they could easily include or represent related species. They would not, however, significantly alter the pattern of distribution apparent in Figure 35. This myctophid is also known from the western North Atlantic Ocean between 13°N and 40°N [Becker, 1967a; Backus et al., 1969; Goode and Bean, 1896; BM uncatalogued, Rosaura 27, 32], from the eastern Atlantic Ocean between 30°N and 35°S, the equatorial Atlantic as

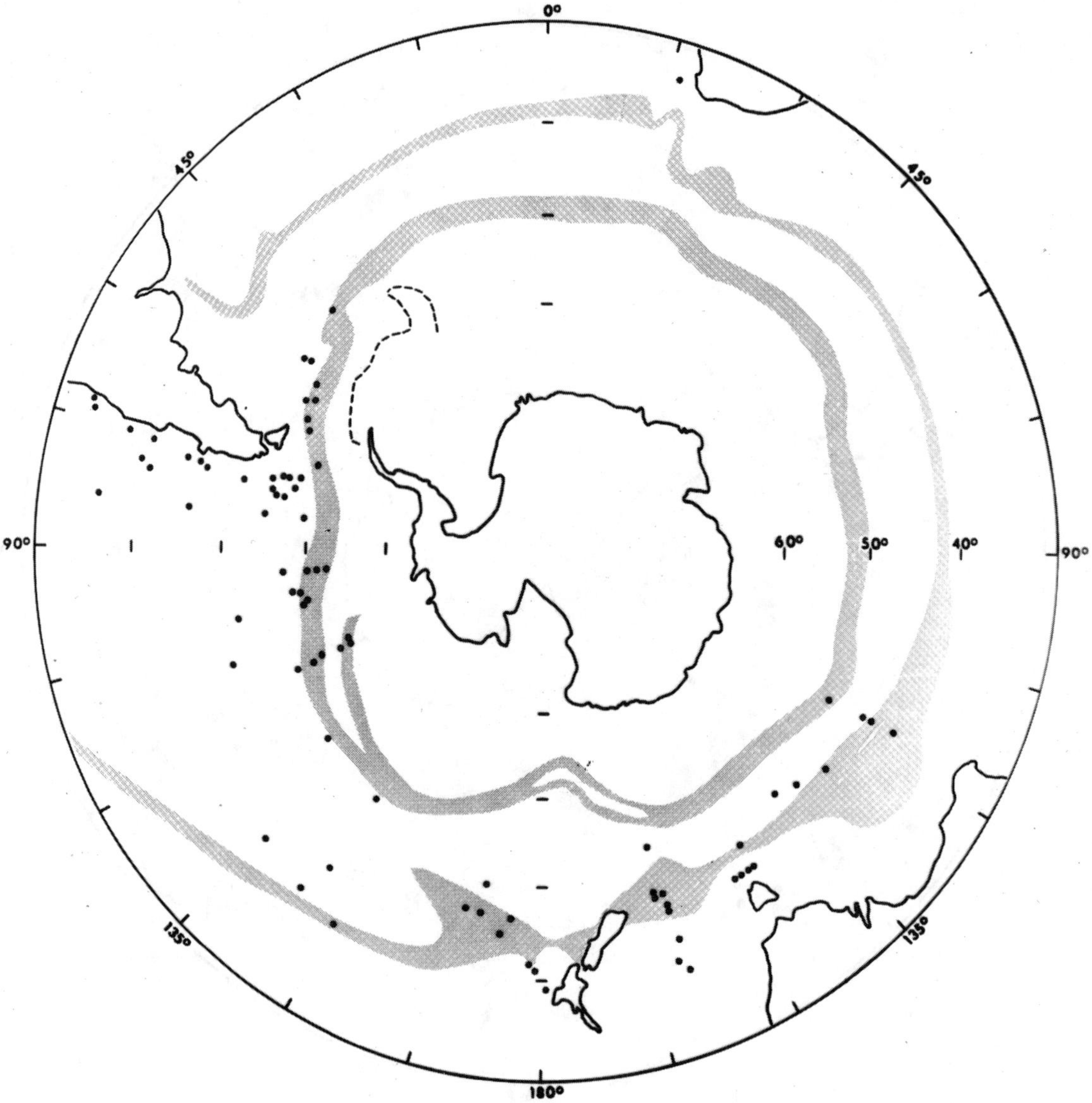

Fig. 36. Distribution of *Lampanyctus macdonaldi* in the southern hemisphere.

far west as 35°W [Backus et al., 1965, 1970; Bolin, 1959; Becker, 1967a; Discovery data], and from between 20°N and 40°N in the western North Pacific [Kulikova, 1960]. It is apparently absent from the Pacific Ocean east of 165°E, the central reaches of the Atlantic subtropical gyres, and the southwestern Atlantic Ocean. The southern limits of distribution of this species between 0° and 160°E, as evident in Figure 35, lie just to the north of the region of the subtropical convergence.

Lampanyctus alatus has often been taken at depths less than 100 m.

Lampanyctus macdonaldi Goode and Bean, 1896
Fig. 36

Lampanyctus intricarius Norman, 1930, p. 330 (part, D 101).
Lampanyctus macdonaldi Bussing, 1965, p. 207 (Drake Passage).--Craddock and Mead, 1970, p. 29 (B13, 5, 59).

Size. *Lampanyctus macdonaldi* attains a length of about 130 mm in the Southern Ocean. Two gravid females were collected by the Eltanin in July.

Distribution. Figure 36 shows the known distribution of L. *macdonaldi* in the southern hemisphere. This species has been collected mainly between the subtropical convergence and the polar front. The absence of records from most of the Atlantic and Indian sectors may be due to inadequate sampling. It is also known from between 37°N and 68°N in the Atlantic Ocean [Bolin, 1959; Becker, 1967a]. Bolin has questioned the validity of a report of L. *macdonaldi* from the Gulf of Mexico by Goode and Bean [1896].

Lampanyctus macdonaldi was captured by the Eltanin only in trawls which fished deeper than 600 m. Juveniles may inhabit somewhat shallower waters than adults do.

Lampanyctus intricarius Tåning, 1928
Fig. 37

Lampanyctus intricarius Norman, 1930, p. 330 (part, D 87, 107, 250).--Bussing, 1965, p. 204

(Elt 80, 190, 742, 743).--Natpaktitis and Natpaktitis, 1969, p. 50 (part, AB 7324).--Craddock
and Mead, 1970, p. 29 (see below).

Size. Lampanyctus intricatius attains a length
of over 170 mm.

Distribution. Figure 37 shows the distribution
of L. intricatius south of 30°S. This species has
been collected in the region of the subtropical
convergence in the Atlantic, Indian, and western
Pacific sectors and south of the convergence in
the Australian and central-eastern Pacific sectors. It is apparently absent from the southwestern Atlantic sector. Although the extensive
material reported by Craddock and Mead [1970] from
the southeastern Pacific almost certainly is L.
intricatius, I have examined the material from
just two of their stations, and only these are
included in Figure 37. This species is also known
from the North Atlantic east of 50°W and between
30°N and 65°N [Bolin, 1959; Becker, 1967a; Discovery data] and from near Bermuda [Beebe, 1937].
The latter record has not been verified.

Some individuals of L. intricatius less than
about 40 mm long have been collected at depths
less than 100 m, whereas specimens larger than 100
mm seem to inhabit waters deeper than 400 m.

Systematic note. Lampanyctus intricatius is
closely related to L. crocodilus, a species found
in the Atlantic north of 22°N [Bolin, 1959] and to
the "Southern Ocean" species L. lepidolychnus
according to Becker [1967a].

Lampanyctus lepidolychnus Becker, 1967

Fig. 38

Lampanyctus intricatius Norman, 1930, p. 330
(part, D 76, 100).--Natpaktitis and Natpaktitis,
1969, p. 50 (part, AB 7072, 7118, 7121, 7324,
7330, 7336).
Lampanyctus lepidolychnus Becker, 1967a, p. 112
(near 42°S, 39°W; near 42°S, 159°E).

Size. Lampanyctus lepidolychnus attains a
length of about 115 mm. It is apparently smaller
than its close relative L. intricatius.

Distribution. Figure 38 shows the distribution

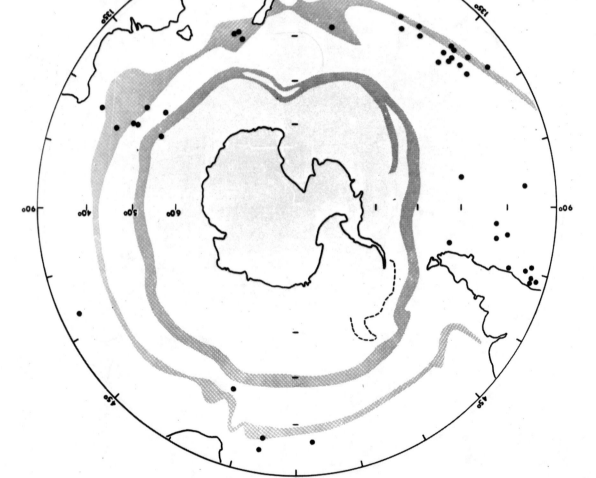

Fig. 37. Distribution of Lampanyctus intricatius south of 30°S.

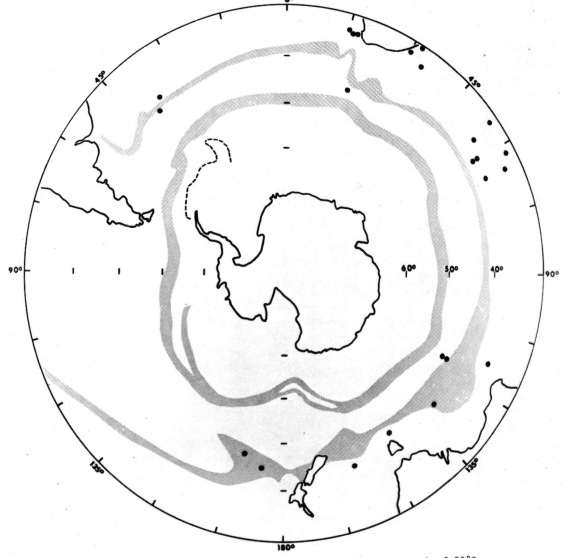

Fig. 38. Distribution of *Lampanyctus lepidolychnus* south of 30°S.

of L. lepidolychnus south of 30°S. It has also been collected near 23°S, 60°E in the Indian Ocean (AB 7072). It is apparently absent in the Pacific sector east of 165°W and seems to be limited to the region in and immediately north of the subtropical convergence.

Juvenile L. lepidolychnus were collected by the Discovery in two trawls which did not fish deeper than 126 m.

Lampanyctus iselinoides Bussing, 1965
Fig. 30

Lampanyctus iselinoides Bussing, 1965, p. 205 (Elt 80, 190, 742, 743).--Craddock and Mead, 1970, p. 29 (B 13, 25 stations; 30°S-34°S, 72°W-80°W).

Size. Lampanyctus iselinoides attains a length of about 100 mm.

Distribution. Figure 30 shows the known distribution of L. iselinoides, which is found between 30°S and 50°S off Chile.

This species has often been taken at depths less than 100 m.

Systematic note. Lampanyctus iselinoides is related to L. crocodilus, L. intricarius, and L. lepidolychnus according to Bussing [1965] and Becker [1967a].

Lampanyctus spp.
Fig. 39

Distribution. The distribution of L. sp. B, L. sp. C, and L. sp. D are shown in Figure 39. Lampanyctus sp. C was collected by the Eltanin at the northernmost stations near the subtropical convergence in the central-eastern Pacific sector. The largest specimen in the collection measured 48 mm long. Lampanyctus sp. D was collected by the Eltanin in and south of the subtropical convergence in the Pacific sector and just north of the convergence in the Tasman and Australian sectors. The record from the southwestern Atlantic may have been mislabeled (Elt 1525). The largest specimen taken was 67 mm long. Gravid females as small as 54 mm were captured in July through January.

Systematic notes. Three forms that are seemingly related to L. festivus [Tåning, 1928] and

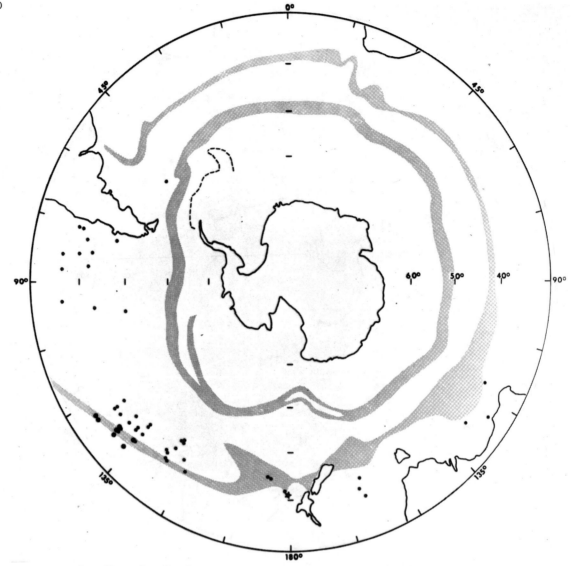

Fig. 39. Distributions of Lampanyctus sp. B (star), Lampanyctus sp. C (large dot), and Lampanyctus sp. D (small dot).

L. tenuiformis [Brauer, 1906] are present in the Eltanin material. A single specimen with high pectoral, dorsal, and anal fin ray counts (17 pectoral, 15 dorsal, and 21 anal) from Eltanin 1401 (L. sp. B) is most similar to L. festivus as described by Nafpaktitis and Nafpaktitis [1969]. However, the VO photophores are straight and not arched as shown in their figure of Tåning's holotype. Two additional forms (L. sp. C and L. sp. D), both well represented in Eltanin collections, are characterized by a lower number of fin rays (14-15 pectoral, 12-13 dorsal, and 16-18 anal) and straight to slightly arched VO photophores. Characters which differentiate L. sp. C from L. sp. D include a more posterior origin of the anal fin relative to the dorsal fin, a well-pigmented supracaudal luminous gland with fused plates, and the presence of a broad vertical band of pigment posterior to the pectoral fin in juveniles less than 30 mm long.

Lobianchia Gatti, 1903

The genus Lobianchia is most closely related to the genus Diaphus [Paxton, 1972]. Bolin [1959] recognized three species of Lobianchia. Only one, L. dofleini, is represented in the study material. Lobianchia urolampa (Gilbert and Cramer), which is known only from near Hawaii, is very distinct and should possibly be placed in a separate genus (B. G. Nafpaktitis, personal communication, 1972). Lobianchia gemellari (Cocco), which is similar to L. dofleini, is known from the Atlantic Ocean between 11°S and 60°N [Nafpaktitis, 1968; Becker, 1967a] and from the Pacific Ocean near 34°S, 90°W [Craddock and Mead, 1970], near the Galapagos Islands [Beebe and Van der Pyl, 1943], and from 16°-32°N, 138°E-175°W [Kulikova, 1961].

Lobianchia dofleini Zugmayer, 1911
Fig. 40

Diaphus dofleini Norman, 1930, p. 332 (part D 250, 251, 254, 285, 286).
Lobianchia dofleini Kulikova, 1961, p. 16 (near 37°S, 180°; near 37°S, 172°E).--Becker, 1967a, p. 98 (near 42°S, 39°W).--Craddock and Mead, 1970, p. 29 (B 13, 15 stations; 30°-34°S, 76°-91°W).

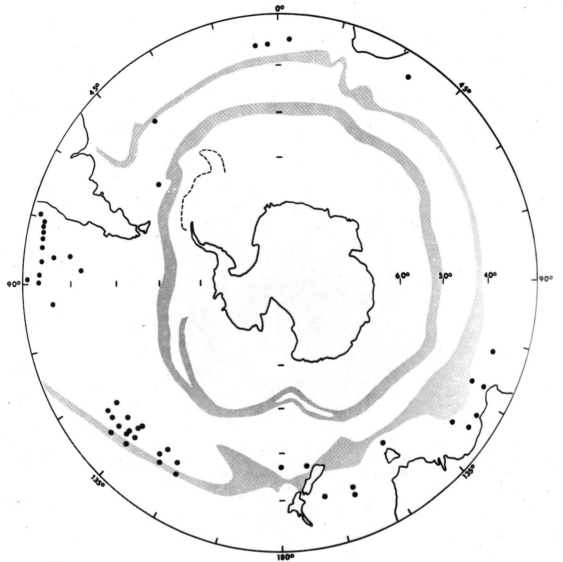

Fig. 40. Distribution of Lobianchia dofleini south of 30°S.

Size. Lobianchia dofleini attains a length of over 45 mm and matures at 32–36 mm [Nafpaktitis, 1968].

Distribution. Figure 40 shows the distribution of L. dofleini south of 30°S. Specimens reported as Myctophum (Diaphus) gemellari from south of 30°S near 90°E by Brauer [1906] represent, or at least include, L. dofleini. This species is known from the North Atlantic Ocean between 25°N and 45°N in the western part and across the middle part to between 50°N and 5°S in the east [Nafpaktitis, 1968]. It has also been taken near 18°S, 4°E in the southwestern Atlantic (D 1602). In the study area, L. dofleini has been collected in and south of the region of the subtropical convergence in the Pacific sector and in and north of the convergence region in the Tasman, Australian, Indian, and Atlantic sectors.

Lobianchia dofleini has often been taken at depths less than 100 m.

Diaphus Eigenmann and Eigenmann, 1890

Approximately 50 nominal species have been estimated to belong to the genus Diaphus (B. G.

Nafpaktitis, personal communication, 1972). Most of these species are restricted to tropical and subtropical waters. Eight species are present in the study material.

Diaphus ostenfeldi Tåning, 1932
Fig. 41

Diaphus ostenfeldi Tåning, 1932, p. 142 (near 35°S, 172°E).—Norman, 1937, p. 86 (BANZARE 71). —Andriashev, 1961, p. 231 (near 42°S, 162°E).— Becker, 1967a, p. 107 (near 42°S, 39°W).— Trunov, 1968, p. 596 (near 32°S, 16°E).—Craddock and Mead, 1970, p. 27 (B 13, 16, 17, 18, 19, 20, 24, 43, 44).

Size. Diaphus ostenfeldi attains a length of over 110 mm. Gravid females as small as 100 mm were collected in August and December.

Distribution. Figure 41 shows the known distribution of D. ostenfeldi south of 30°S. This species is known from one locality north of 30°S, in the Benguela Current near 24°S, 12°E [Trunov, 1968]. Diaphus ostenfeldi has been collected in and north of the region of the subtropical con-

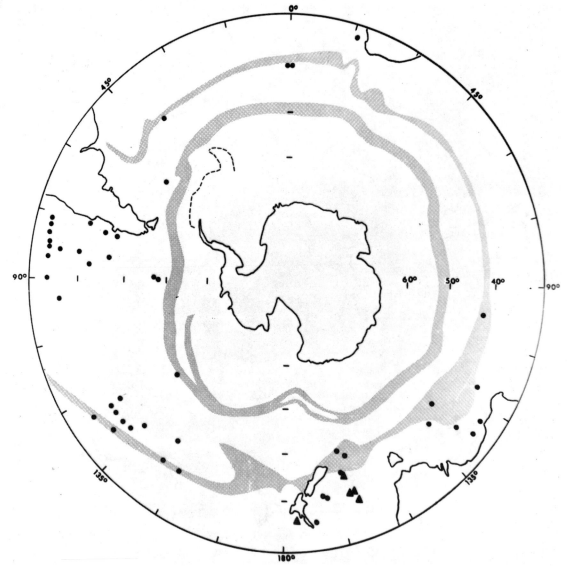

Fig. 41. Distribution of <u>Diaphus ostenfeldi</u> south of 30°S (dot) and of <u>Diaphus danae</u> (triangle).

vergence in the Tasman, Australian, Indian, and Atlantic sectors and in and south of the convergence in the Pacific sector. Four large individuals (all larger than 57 mm) were captured near the Polar Front in the central-eastern Pacific and southwestern Atlantic sectors.

Systematic note. According to Tåning [1932], <u>D. ostenfeldi</u> is quite distinct from its congeners.

Diaphus effulgens Goode and Bean, 1896
Fig. 42

<u>Diaphus effulgens</u> Craddock and Mead, 1970, p. 27 (B 13, 26, 28).

Size. <u>Diaphus effulgens</u> reaches a length of over 120 mm in the North Atlantic Ocean [Nafpaktitis, 1968]. The largest specimen in the study collections from the southern hemisphere measured 82 mm (Elt 1764).

Distribution. Figure 42 shows the known distribution of <u>D. effulgens</u> in the southern hemisphere. In addition to the records shown in the figure, this species is known from the Atlantic Ocean between 18°N and 45°N [Nafpaktitis, 1968].

Diaphus danae Tåning, 1932
Fig. 41

<u>Diaphus danae</u> Tåning, 1932, p. 140 (near 36°S, 176°E).

Size. <u>Diaphus danae</u> attains a length of over 120 mm [Tåning, 1932].

Distribution. Figure 41 shows the known distribution of <u>D. danae</u>. This species has been collected near New Zealand and in the Tasman Sea.

Diaphus mollis Tåning, 1928

Size. <u>Diaphus mollis</u> attains a length of about 60 mm and matures at about 30 mm in the North Atlantic [Nafpaktitis, 1968].

Distribution. This species is known from only one locality (Elt 2223) near 38°S, 100°E in the

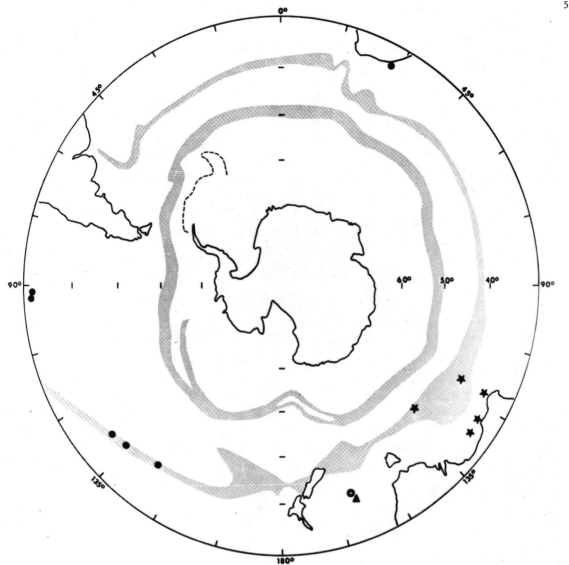

Fig. 42. Distributions of Diaphus effulgens (dot), Diaphus sp. B (star), Diaphus mollis (triangle), and Diaphus termophilus (black dot with white center).

southern hemisphere. It is also known from the North Atlantic between 2°S and 50°N [Nafpaktitis, 1968; Becker, 1967a].

Diaphus termophilus Tåning, 1928

Size. D. termophilus attains a length of about 75 mm and matures at a size larger than 56 mm in the North Atlantic [Nafpaktitis, 1968].

Distribution. Diaphus termophilus is known from only one locality (Elt 1821) near 47°S, 160°E in the southern hemisphere. This species is also known from the North Atlantic west of 39°W, where it appears to be largely confined to waters in and near the Caribbean Sea [Nafpaktitis, 1968]. Becker [1967a] alludes to material of this or closely related species from the Pacific.

Diaphus parri Tåning, 1932
Fig. 43

Size. Diaphus parri attains a length of at least 55 mm.

Distribution. The distribution of D. parri south of 30°S is shown in Figure 43. This species may be represented in the Diaphus theta 'complex' reported by Craddock and Mead [1970] from the southeastern Pacific Ocean, especially their westernmost stations. Diaphus parri is also known from one locality north of 30°S, at about 27°S, 175°E [Tåning, 1932]. The southern limits of this species are in the region of the subtropical convergence and overlap the northern limits of distribution of D. sp. A. It is apparently absent from the southwestern Atlantic Ocean.

This lanternfish has often been collected at depths less than 100 m.

Diaphus sp. A
Fig. 43

Diaphus theta Bussing, 1965, p. 203 (Elt 80, 742, 743).--Becker, 1967a, p. 102 (part, Tasman Sea, near 42°S, 39°W).

54

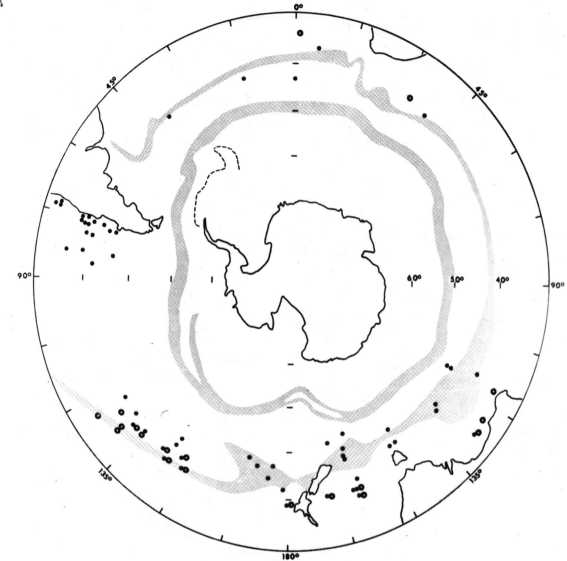

Fig. 43. Distributions of Diaphus parri (black dot with white center)
and Diaphus sp. A (dot) south of 39°S.

Size. Diaphus sp. A attains a length of over
75 mm.

Distribution. Figure 43 shows the distribution
of D. sp. A south of 30°S. The D. theta recorded
by Becker [1967a] from near 39°S, 68°E and the D.
theta 'complex' reported by Craddock and Mead
[1970] from the southeastern Pacific are not included in the figure. The latter material undoubtedly included D. sp. A. The D. theta reported by Bussing [1965] from near 24°S off Chile
is D. sp. A.

This species has often been collected at depths
less than 100 m.

Systematic note. Diaphus sp. A is very similar
to and possible conspecific with D. theta, a species known from subarctic and transitional waters
of the North Pacific Ocean. Diaphus sp. A is
similar to D. parri.

Diaphus sp. B
Fig. 42

Diaphus sp. B is an undescribed form closely
related to the North Atlantic species D. rafines-

qui (B. G. Nafpaktitis, personal communication,
1972). This species has been collected only south
of western Australia (Figure 42).

Hintonia Fraser-Brunner, 1949

The genus Hintonia includes one known species,
H. candens.

Hintonia candens Fraser-Brunner, 1949
Fig. 44

Hintonia candens Fraser-Brunner, 1949, p. 1104
(D 1774).

Size. Hintonia candens attains a length of at
least 115 mm.

Distribution. The known distribution of H.
candens is shown in Figure 44. This species has
been collected in the region of the subtropical
convergence in the Atlantic, Australian, and Tasman sectors and in, as well as south of, the convergence between 120°W and 170°W in the Pacific

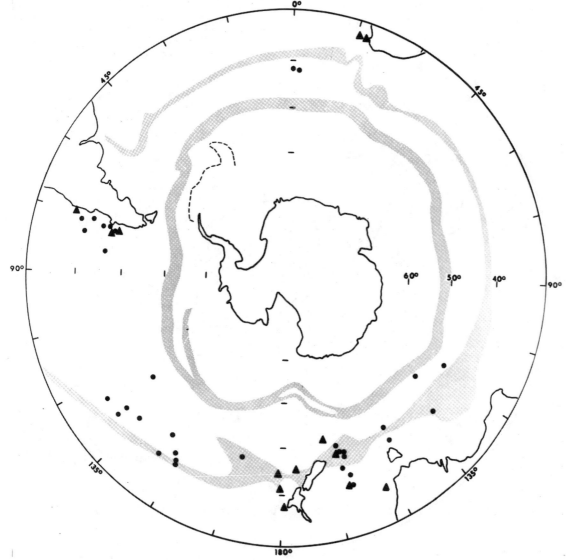

Fig. 44. Distributions of _Hintonia candens_ (dot) and _Lampanyctodes hectoris_ (triangle).

sector. It is also known from between 40°S and 50°S off Chile.

This myctophid was taken in only one trawl which did not fish as deep as 200 m (WS 588, 0-100 m).

Lampanyctodes Fraser-Brunner, 1949

The genus _Lampanyctodes_ includes one known species, _L. hectoris_.

Lampanyctodes hectoris (Günther, 1864)
Fig. 44

Scopelus hectoris Günther, 1876, p. 399 (Cook Strait, New Zealand).
Scopelus argenteus Gilchrist, 1904, p. 15 (near South Africa).
Lampanectus hectoris Norman, 1930, p. 328 (D 99d).

Size. _Lampanyctodes hectoris_ attains a length of over 55 mm.
Distribution. The known distribution of _L._

hectoris is shown in Figure 44. This species is primarily distributed in the region of the subtropical convergence off Chile, New Zealand, Australia, and South Africa. It is apparently absent in the waters of the southwestern Atlantic sector.

Some _L. hectoris_ have been collected at the surface. The fact that a number of _L. hectoris_ have been collected in bottom trawls and that the species is primarily found near land masses may indicate that it is limited to continental shelf and slope waters.

Scopelopsis Brauer, 1960

The genus _Scopelopsis_ includes one known species, _S. multipunctatus_.

Scopelopsis multipunctatus Brauer, 1906
Fig. 45

Scopelopsis multipunctatus Brauer, 1906, p. 146 (near 33°S, 16°E).--Barnard, 1925, p. 246 (near South Africa).--Norman, 1930, p. 318 (D 101).

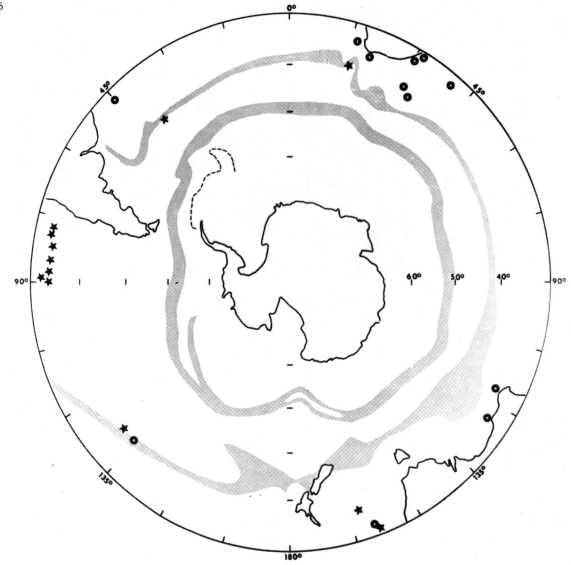

Fig. 45. Distributions of Scopelopsis multipunctatus (black dot with white center) and Notoscopelus resplendens (star) south of 30°S.

Scopelopsis caudalis Whitley, 1932, p. 333 (Lord Howe Island).

Size. Scopelopsis multipunctatus attains a length of about 75 mm and spawns in July and August [Legand, 1967].

Distribution. The known distribution of S. multipunctatus south of 30°S is shown in Figure 45. This species is also known from north of the Kermadec Islands [Becker, 1967b], from between 14°S and 23°S in the eastern Pacific (B. Davy, personal communication, 1973), from as far north as 10°S in the eastern Indian Ocean [Legand, 1967] and 23°S in the western central Indian Ocean [Nafpaktitis and Nafpaktitis, 1969], and from two localities in the Atlantic Ocean near 28°S, 43°W (D 712) and 13°N, 24°W (D 2073). Except for the latter record, which is probably erroneously labeled, S. multipunctatus appears to be limited to subtropical waters of the southern hemisphere. It has been taken farther north in the eastern reaches of the Pacific and Indian oceans than in the western reaches.

Notoscopelus Günther, 1864

The genus Notoscopelus comprises at least five species (B. G. Nafpaktitis, personal communication, 1974). Notoscopelus kroyeri (Malm) is restricted to cold waters of the North Atlantic Ocean, and N. japonicus Tanaka is restricted to the northwestern Pacific Ocean. The remainder appear to be distributed primarily in warm waters of the World Ocean. One of these species, N. resplendens, is represented in the study collections.

Notoscopelus resplendens (Richardson, 1844)
Fig. 45

Notoscopelus ejectus Waite, 1904, p. 150 (Lord Howe Island, see Bolin, 1959, p. 40).
Notoscopelus resplendens Becker, 1967a, p. 122 (near 42°S, 39°W).--Craddock and Mead, 1970, p. 30 (B 13 16, 17, 18, 19, 20, 21, 22, 23, 24, 30, 43).

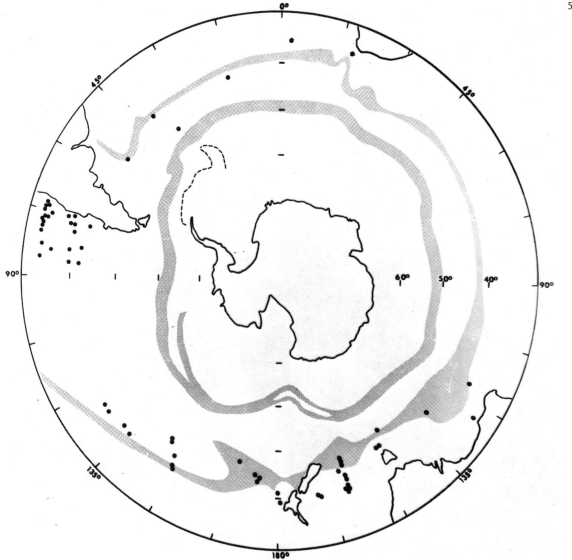

Fig. 46. Distributions of *Lampichthys procerus* south of 30°S.

Figure 45 shows the distribution of N. resplendens south of 30°S. The *Lampanyctus elongatus* reported by Norman [1930] from near 61°S, 48°W and shown by Andriashev [1962] to be N. resplendens is not included in the figure because of a discrepancy between the depth data published with that station (D 168) and the data with the specimen as reported by Norman. The *Lampanyctus elongatus* and *Myctophum (Lampanyctus) elongatum* reported by Lütken [1892] and Barnard [1925], respectively, from near South Africa could actually be N. resplendens. This species is also known from the Atlantic Ocean south of 38°N [Bolin, 1959], from the Indian Ocean between 24°S and 29°S [Nafpaktitis and Nafpaktitis, 1969], from the northwestern Pacific Ocean [Mead and Taylor, 1953], from the northeastern Pacific Ocean near 35°N [Berry and Perkins, 1966], and from larvae in the eastern tropical Pacific Ocean [Ahlstrom, 1971]. The southern limits of the distribution of N. resplendens correspond quite closely to the region of the subtropical convergence.

Lampichthys Fraser-Brunner, 1949

The genus *Lampichthys* includes one known species, L. procerus (Paxton, 1972).

Lampichthys procerus (Brauer, 1904)
Fig. 46

Myctophum (Lampanyctus) procerum Brauer, 1904, p. 402 (near 35°S, 18°E).--Pappenheim, 1914, p. 195 (near 35°S, 2°E).
Lampichthys rectangularis Fraser-Brunner, 1949, p. 1103 (D 717).--Bussing, 1965, p. 207 (Elt 190, 743).--Becker, 1967a, p. 121 (near 42°S, 39°W).--Craddock and Mead, 1970, p. 29 (B 13, 16 stations; 30°-34°S, 73°-84°W).

Size. *Lampichthys procerus* attains a length of over 80 mm.
Distribution. The known distribution of L. procerus south of 30°S is shown in Figure 46. This species is also known from one locality in

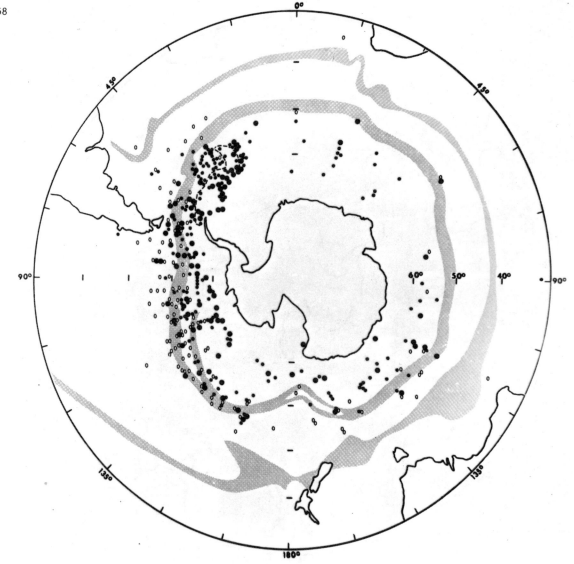

Fig. 47. Distribution of *Gymnoscopelus braueri*. Small dot indicates individuals 40-100 mm long, open circle indicates sample includes individuals less than 40 mm long, and large dot indicates sample includes individuals larger than 100 mm long.

the Peru Current at about 18°S [Bussing, 1965]. It has been collected primarily in the region of the subtropical convergence and between 18°S and 44°S off the west coast of South America. Its absence in the Indian sector may be due to inadequate sampling.

A few specimens of L. procerus have been collected at depths less than 100 m.

Gymnoscopelus Günther 1873

The genus *Gymnoscopelus* includes nine species currently placed in two subgenera. All species, which are essentially limited to Antarctic and subantarctic waters, are represented in the study material.

Subgenus Gymnoscopelus Günther, 1873

The genus *Gymnoscopelus* includes two species pairs: G. braueri-G. opisthopterus and G. aphya-

G. bolini, both pairs are essentially limited to Antarctic waters.

Gymnoscopelus braueri (Lönnberg, 1905)
Fig. 47

Mycotophum (*Lampanyctus*) *braueri* Lönnberg, 1905, p. 64 (near 49°S, 51°W).
Lampanyctus braueri Norman, 1930, p. 327 (part, D 62, 66, 239).
Gymnoscopelus braueri Andriashev, 1962, p. 264 (OB 48, 368, 409, 455).

Size. *Gymnoscopelus braueri* attains a length of about 120 mm. Gravid females as small as 100 mm long were collected in June and July.

Distribution. Figure 47 shows the known distribution of G. braueri. Two of the specimens reported as *Lampanyctus braueri* by Norman [1937] were examined and found to be G. opisthopterus. Should the rest of Norman's specimens, which I

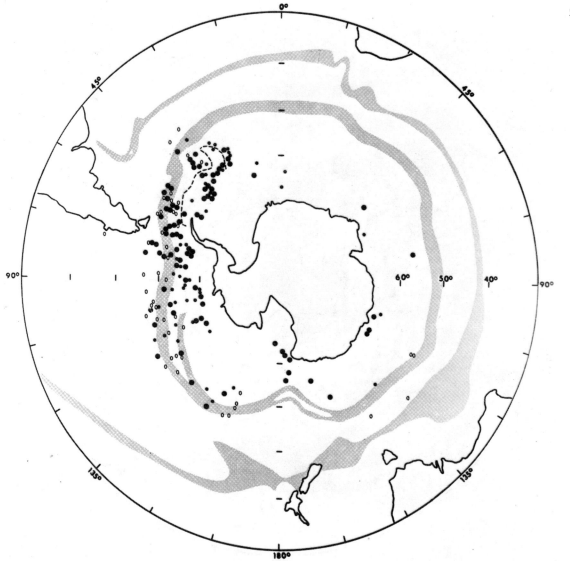

Fig. 48. Distribution of *Gymnoscopelus opisthopterus*. Small dot indicates individuals 45-100 mm long, open circle indicates sample includes individuals less than 45 mm long, and large dot indicates sample includes individuals larger than 100 mm long.

did not see, actually be G. braueri, the distribution would be extended to 65°S in the Indian sector. An unexamined and unplotted specimen reported by Regan [1913] would, if it is actually G. braueri, extend the distribution of this species to the northeastern edge of the Weddell Sea. A number of records of G. braueri from 60°-70°S [Andriashev, 1962; Slava material] that are not included in Figure 47 because of lack of information on the length of the specimens would not alter the apparent distribution. Whitley described G. piabilis from two of three specimens from Macquarie Island reported by Waite [1916] under the name Lampanyctus braueri. Should the third specimen actually be G. braueri, it would have no effect on the distribution shown in Figure 47.

Juvenile G. braueri have been collected farther north than, and not as far south as, adults. With few exceptions, individuals less than 40 mm long and more than 100 mm have not been taken south or north of the region of the Polar Front, respectively. Most of the northernmost records are of juveniles less than 40 mm. All individuals captured more than 2° of latitude south of the Weddell-Scotia confluence were larger than 65 mm.

Numerous G. braueri, all less than 100 mm, have been collected at depths less than 100 m during the night. Larger individuals have been collected in trawls which have fished deeper than 100 m.

Gymnoscopelus opisthopterus Fraser-Brunner, 1949
Fig. 48

Lampanyctus braueri Norman, 1930, p. 327 (part, D 151); 1937, p. 86 (part, BANZARE 32, 96).
Gymnoscopelus opisthopterus Fraser-Brunner, 1949, p. 1102 (D 1718).--Andriashev, 1962, p. 268 (Slava 73).

Size. *Gymnoscopelus opisthopterus* attains a length of over 160 mm. Gravid females 142 mm long were collected from August to January.

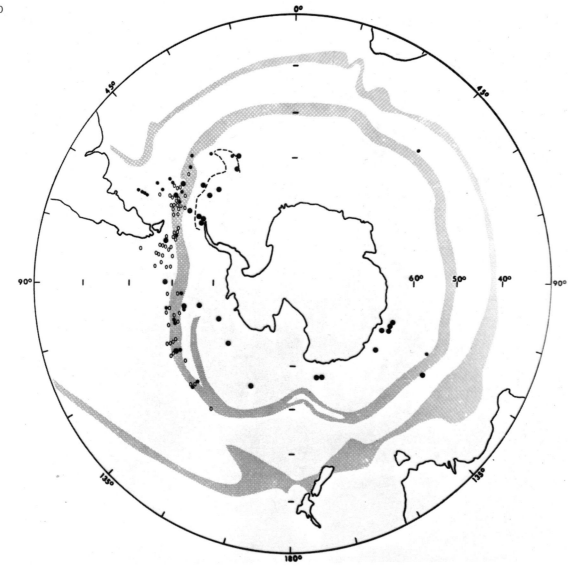

Fig. 49. Distribution of Gymnoscopelus aphya. Open circles indicate individuals less than 35 mm long, small dot indicates individuals 35-79 mm long, and large dots indicate individuals larger than 99 mm long.

Distribution. Figure 48 shows the known distribution of G. opisthopterus. Juveniles have been collected farther north than and not as far south as larger individuals. With few exceptions, individuals less than 45 mm or greater than 100 mm have not been taken south or north of the region of the Polar Front, respectively. Juveniles are distributed primarily in the region of the Polar Front, whereas adults occur throughout Antarctic waters except in the southern and central regions of the Ross and Weddell seas. The species appears to be absent in northern Antarctic waters of the eastern Atlantic and Indian sectors. The northernmost records for the species, south of Australia, between Chile and 105°W and near 45°W are those of juveniles less than 45 mm long.

Gymnoscopelus opisthopterus has been taken mainly below 500 m. Two juveniles less than 40 mm long were captured at 66 m, the shallowest depth.

Gymnoscopelus aphya Günther, 1873
Fig. 49

Gymnoscopelus aphya Günther, 1873, p. 267 (near 55°S, 85°W).
Lampanyctus nicholsi Gilbert, 1911, p. 17 (near 47°S, 60°W).--Norman, 1930, p. 326 (part, D 60, 62; WS 236).
Gymnoscopelus nicholsi Andriashev, 1962, p. 270 (Slava 57, 61, 72, 75, 76, 78, 85, 103, 108).

Size. Gymnoscopelus aphya attains a length of at least 150 mm.

Distribution. The distribution of G. aphya is shown in Figure 49. Angelescu and Cousseau [1969] reported and figured additional material taken from the stomach of hake collected on the Argentine Shelf. Juvenile G. aphya have been collected farther north than, and not as far south as, large individuals. Individuals less than 34 mm have

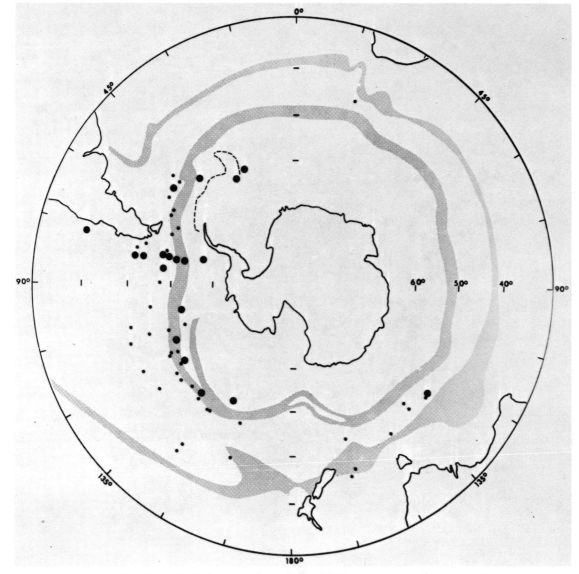

Fig. 50. Distribution of *Gymnoscopelus bolini*. Small dot indicates individuals less than 100 mm long, and large dot indicates individuals larger than 99 mm long.

been captured in and north of the Polar Front between 45°W and 145°W. Individuals 35–79 mm long have been taken near the Polar Front in the Indian and Pacific sectors as well as north and south of the front in the southwestern Atlantic sector. Specimens larger than 98 mm have been collected from near the Polar Front to south of the Weddell-Scotia confluence in the southwestern Atlantic and to near the Antarctic continent in the Pacific and Australian sectors.

Numerous juveniles and one individual 150 mm long have been caught at depths less than 100 m.

Gymnoscopelus bolini Andriashev, 1962
Fig. 50

Lampanyctus *nicholsi* Norman, 1930, p. 326 (part, D 107, 217).
Gymnoscopelus *bolini* Andriashev, 1962, p. 273 (OB 394, 416).

Size. *Gymnoscopelus bolini* attains a length of over 240 mm.

Distribution. Figure 50 shows the known distribution of *G. bolini*. Except off Chile, individuals less than 100 mm long have been collected farther north than, and not as far south as, larger individuals, which were taken mainly in the region of the Polar Front between about 60°W and 160°W near the eastern edge of the Weddell-Scotia confluence and as far north as 42°S off Chile. Juveniles have been caught in and near the subtropical convergence south of Australia and New Zealand but well to the south of the convergence in the eastern and central parts of the Pacific sector. The paucity of records from most of the Atlantic and Indian sectors could be due to insufficient sampling.

Juvenile *G. bolini* less than 100 mm long have often been captured at depths less than 100 m. Larger individuals live deeper than 450 m.

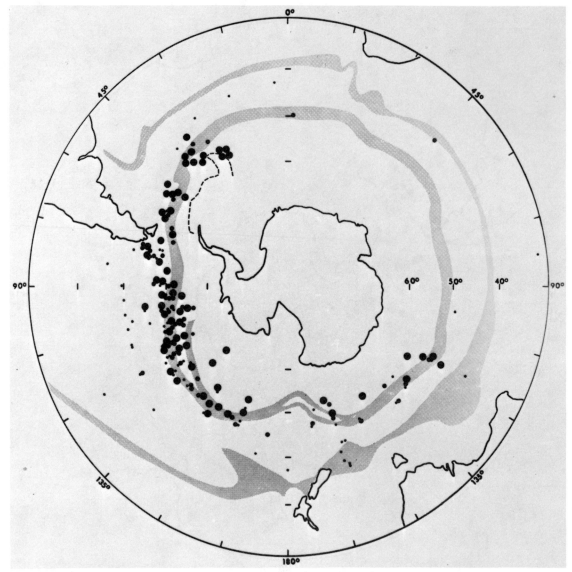

Fig. 51. Distribution of Gymnoscopelus (Nasolychnus) fraseri. Small dot indicates individuals less than 45 mm long, medium dot indicates individuals 45-64 mm long, and large dots indicates individuals larger than 64 mm long.

Nasolychnus Smith, 1933

The subgenus Nasolychnus includes five species, three of which are undescribed and all of which are represented in the study collections.

Gymnoscopelus piabilis and G. sp. C are related and can be distinguished from other Nasolychnus species by the presence of more than 30 gill rakers on the first gill arch (very rarely less), the origin of the dorsal fin over the origin of the ventral fin, and the presence of well-developed luminous tissue anterior to the lower half of the eye. Gymnoscopelus fraseri and G. sp. A are very closely related and can be distinguished from other Nasolychnus species by the presence of fewer than 29 gill rakers on the first gill arch (very rarely more), the origin of the dorsal fin slightly in advance of the ventral fin, and, with the exception of male G. sp. A, the absence of well-developed luminous tissue anterior to the lower half of the eye. Gymno-

scopelus sp. B is distinguished by the origin of the dorsal fin distinctly in advance of the ventral fin and the presence of a continuous band of luminous tissue anterior to the eye.

Gymnoscopelus (Nasolychnus) fraseri
Fraser-Brunner, 1931
Fig. 51

Size. Gymnoscopelus fraseri attains a length of about 80 mm. Gravid females about 75 mm long were collected from June to August.

Distribution. Figure 51 shows the distribution of G. fraseri south of 30°S. The specimens reported as Gymnoscopelus (Nasolychnus) sp. by Andriashev [1962] almost certainly include G. fraseri. Gymnoscopelus fraseri has also apparently been collected near 3°S, 5°E [Fraser-Brunner, 1931] and near 24°N, 21°W (D 2067). These records are puzzling and represent the only records for the genus from north of the Southern Ocean. Ju-

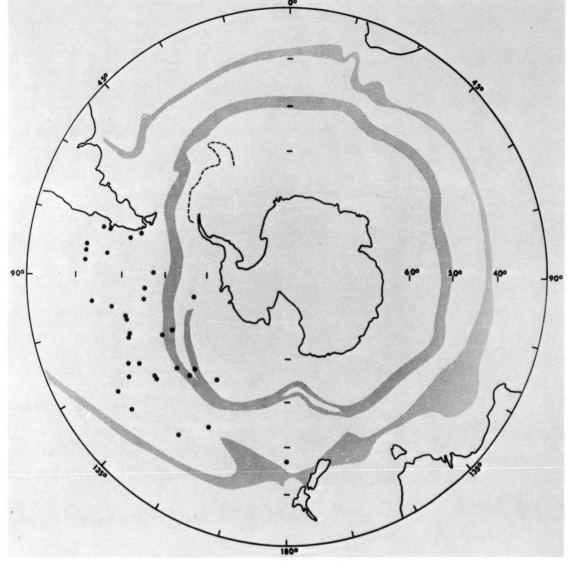

Fig. 52. Distribution of <u>Gymnoscopelus</u> (<u>Nasolynchnus</u>) sp. A.

venile <u>G</u>. <u>fraseri</u> have been taken farther north than, and not as far south as, adults. Individuals less than about 45 mm long have been collected primarily between the Polar Front and 5°-10° north of the front. Individuals larger than about 65 mm have been captured primarily in the waters of the Polar Front. A few have been taken farther north, between 105°W and Chile.

<u>Gymnoscopelus</u> <u>fraseri</u> has repeatedly been collected at depths less than 100 m.

<u>Systematic notes</u>. <u>Gymnoscopelus</u> <u>fraseri</u> may be distinguished from the closely related <u>G</u>. sp. A by the absence of antorbital luminous tissue anterior to the lower half of the eye and by a higher number of rakers on the first gill arch (26-28 versus 24-26). Many juveniles in the collections examined could not be assigned with certainty to either species because of undeveloped antorbital luminous tissue.

<u>Gymnoscopelus</u> (<u>Nasolychnus</u>) sp. A
Fig. 52

<u>Size</u>. <u>Gymnoscopelus</u> sp. A attains a length of at least 60 mm.

<u>Distribution</u>. Figure 52 shows the known distribution of <u>G</u>. sp. A. It has been collected only in the Pacific sector between the Polar Front and 40°-45°S, where it seems to be endemic.

This species has been collected at depths less than 100 m.

<u>Gymnoscopelus</u> (<u>Nasolychnus</u>) sp. B
Fig. 53

<u>Lampanyctus</u> <u>braueri</u> Norman, 1930, p. 327 (part, D 257).

<u>Gymnoscopelus</u> <u>piabilis</u> Andriashev, 1962, p. 275 (Slava 73, 103, 108).

<u>Size</u>. <u>Gymnoscopelus</u> sp. B attains a length of at least 110 mm.

<u>Distribution</u>. Figure 53 shows the known distribution of <u>G</u>. sp. B. Juveniles have been collected farther north than, and not as far south as, adults. Individuals less than about 60 mm long have been found between the Polar Front and the region of the subtropical convergence. The northernmost records are of specimens less than

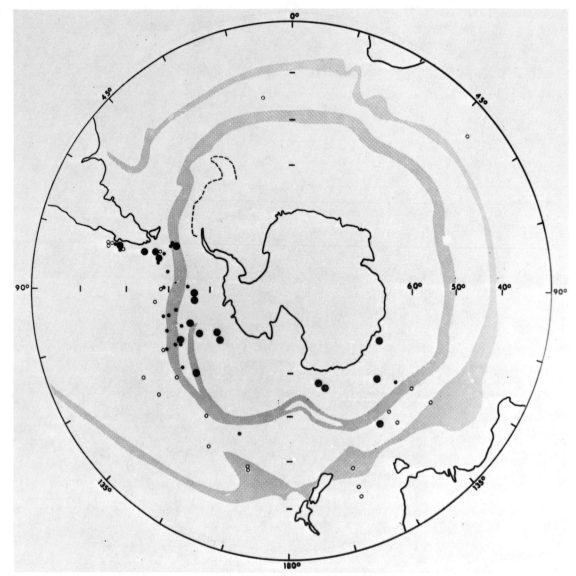

Fig. 53. Distribution of <u>Gymnoscopelus</u> (<u>Nasolychnus</u>) sp. B. Open circle
indicates individuals less than 61 mm long, small dot indicates individuals 61-94 mm long, and large dots indicates individuals larger than 94
mm long.

35 mm in length. Specimens 61-94 mm long have
been taken near the Polar Front and to about 47°S
off Chile. Larger individuals were caught south
of the Polar Front to the region of the Antarctic
divergence in the Pacific and Australian sectors
and to about 47°S off Chile. Inadequate sampling
could explain the absence of this species from
most of the Atlantic and Indian sectors. It appears to be absent from the eastern reaches of
the Drake Passage.

Specimens less than about 90 mm long have been
collected at depths less than 100 m, whereas larger individuals have only been taken in trawls
which fished deeper than 500 m.

Gymnoscopelus (Nasolychnus) piabilis
(Whitley, 1931)
Fig. 54

Lampanyctus piabilis Whitley, 1931, p. 103
    (Macquarie Island).
Myctophum (Nasolychnus) florentii Smith, 1933,
    p. 126 (off South Africa).

Size. <u>Gymnoscopelus piabilis</u> attains a length
of over 110 mm. The holotype of <u>G. florentii</u> is
the only known gravid female.

Distribution. The known distribution of <u>G.
piabilis</u> is shown in Figure 54. This species has
been primarily collected in subantarctic waters
of the Atlantic and Australian sectors and between
40°S and 55°S off Chile. It has also been taken
off South Africa. Its absence from the Indian
sector may be due to inadequate sampling. It
appears to be absent from most of the Pacific
sector.

Systematic notes. <u>Gymnoscopelus piabilis</u> can
be distinguished from the closely related <u>G.</u> sp.
C by the presence of dense pigment on the distal
half of the ventral fins, a narrower adipose fin
base (less than versus more than two photophore
diameters), and a narrower lateral line canal
(less than versus more than two photophore diameters). Although I have not examined the holotype
of <u>Myctophum</u> (<u>Nasolychnus</u>) <u>florentii</u>, there is
evidence in support of Bolin's [Andriashev, 1962,
p. 274] contention that it is synonymous with <u>G.</u>

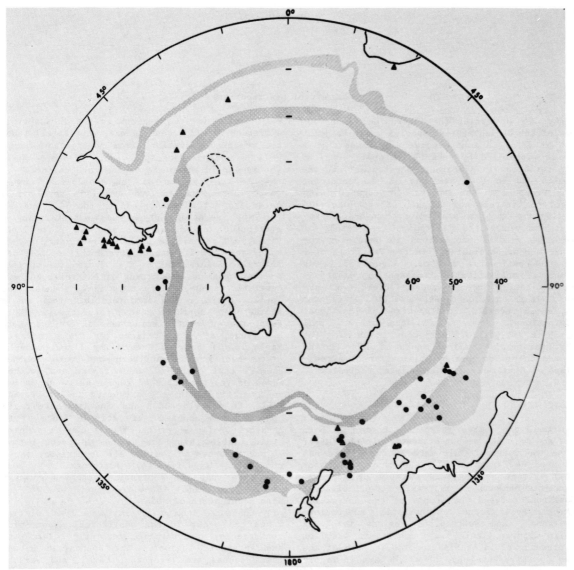

Fig. 54. Distributions of <u>Gymnoscopelus</u> (<u>Nasolynchnus</u>) <u>piabilis</u> (triangle) and <u>Gymnoscopelus</u> (<u>Nasolynchnus</u>) sp. C (dot).

<u>piabilis</u>. The description and figure of the holotype [Smith, 1933] indicate that the distal halves of the ventral fins are darker than the proximal halves, that the adipose fin is small, and that the lateral line is relatively narrow.

Gymnoscopelus (Nasolychnus) sp. C
Fig. 54

<u>Gymnoscopelus</u> sp. Nafpaktitis and Nafpaktitis, 1969, p. 66 (AB 1727).

<u>Size</u>. <u>Gymnoscopelus</u> sp. C attains a length of at least 110 mm.

<u>Distribution</u>. Figure 54 shows the known distribution of <u>G</u>. sp. C. This species has been collected in subantarctic waters of the Australian, western Pacific, and Indian sectors. Inadequate sampling may account for the lack of records in most of the Atlantic and Indian sectors, whereas this species appears to be absent from off Chile and from subantarctic waters away from the Polar Front in the central-eastern Pacific Sector.

Figure 55 summarizes the latitudinal distributions of the 84 lanternfish species known to occur south of 30°S. A more precise diagrammatic summary is prohibited by several factors: longitudinal variation in hydrology, particularly in the dynamics and in the positions of the Antarctic Polar Front and subtropical convergence with respect to latitude and to each other; localized and disjunct species populations; anomalous records; and restricted latitudinal extensions of individual species populations in unique hydrographic areas, particularly in the Pacific sector and near continental margins. Figure 55 is an attempt at realistically representing distributions by taking some of these factors into account. It is based on close inspection of Figures 3-54, which reveal a broad spectrum of individual distributions. However, it does allow formal geographic analysis. The species in each of the major hydrographic regions south of 30°S can be compared to one another by utilization of a coefficient of difference (CD) calculated as follows:

$$CD = (1 - C/N) \times 100$$

where C is the number of species common to both areas and N is the number of species in the larger of the two faunas. This method of analysis has been used by Savage [1960], who recognized major faunal differences where CD was 50% or greater. A convenient means of expressing the individual comparisons between the faunas of the Antarctic, Antarctic Polar Front, subantarctic, subtropical convergence, and subtropical regions is by a trellis diagram (Table 1). Numbers are based on the presence of laternfish species in each of the areas as represented by the solid bars in Figure 55.

Table 1a indicates that, with the possible exception of the subantarctic region, the fauna of

each of the various regions is not substantially different from adjacent faunas, whereas all faunas are remarkably distinct from nonadjacent faunas. This is not unexpected, however, if one considers the manner in which the regions are defined. For example, the region of the subtropical convergence, as emphasized previously, is merely indicated and not circumscribed by the 34.6 and 34.8 per mil isohalines at 200 m; subantarctic and subtropical waters do not stop at these isohalines, but rather are modified by mixing, and the mixed area itself extends beyond this region. The Antarctic Polar Front is also dynamic, with latitudinal movement and mixing of Antarctic and subantarctic waters, and is not a fixed boundary isolating Antarctic and subantarctic waters as its southern and northern edges, respectively. Moreover, the apparent distinctness of the subantarctic fauna, as it is defined for the purposes of Table 1b, is as much due to the large number of predominantly more northern and more southern species that occur at the northern and southern edges of the subantarctic region as to an endemic subantarctic fauna. By extending the subtropical region to north of 30°S and the Antarctic region to more than about 5° of latitude south of the Antarctic Polar Front the faunas can be compared again (Table 1b). The lowest CD value between adjacent faunas is 52% and all comparisons can be considered significantly different. I would emphasize that the subtropical species shown in Figure 55 are only those from south of 30°S that also occur north of 30°S.

I conclude from this analysis that distinct faunas do occur in the subtropical, subtropical convergence, subantarctic, Antarctic Polar Front, and antarctic regions. The boundaries between these faunas are irregular, mixed, and variable and are approximated by the Antarctic Polar Front and a region of rapid change in salinity at 200 m (i.e., the subtropical convergence).

Table 1a. Number of Species in Common and Coefficient of Difference Between Lanternfish Faunas of Major Hydrographic Regions South of 30°S

	ST	STC	SA	APF	A
ST	54	19%	83%	100%	100%
STC	44	53	70%	94%	98%
SA	9	16	27	52%	70%
APF	0	3	13	19	32%
A	0	1	8	13	13

Numbers are based on the presence or absence of species as determined by the solid bars in Figure 55. Abbreviations are as follows: A, Antarctic, APF, Antarctic Polar Front; SA, subantarctic; STC, subtropical convergence; and ST, subtropical.

Table 1b. Number of Species in Common and Coefficient of Difference Between Lanternfish Faunas of Major Hydrographic Regions South of 30°S

	30°S	STC	SA	APF	CA
<30°S	28	62%	100%	100%	100%
STC	20	53	70%	94%	98%
SA	0	16	27	52%	96%
APF	0	3	13	19	58%
CA	0	1	1	8	8

This table is the same as Table 1a except that the subtropical region is restricted to less than 30°S (<30°S), and the Antarctic region is restricted to more than 5° latitude south of the Antarctic Polar Front (CA). See text.

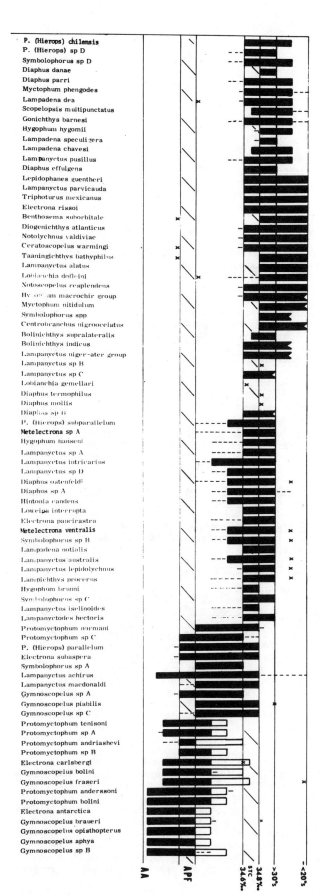

Fig. 55. Diagrammatic summary of the latitudinal distribution of 84 lanternfish species known to occur south of 30°S. A solid bar represents the usual latitudinal extent of a species, an open bar represents the usual presence of juveniles in areas where adults of a particular species are generally excluded, a dashed line represents the occurrence of individual species in restricted longitudes, and an X represents isolated or anomalous records. AA denotes Antarctic continent; APF, Antarctic Polar Front; and STC, subtropical convergence as delineated by the 34.6 and 34.8 per mil isolhalines at 200 m.

Patterns of Distribution

Five patterns of distributuion are apparent in Figure 55: a group of species with northern limits of distribution generally north of 30°S and southern limits of distribution generally in or very near the subtropical convergence; a group of species with northern limits of distribution generally between 30°S and the subtropical convergence and with southern limits of distribution generally less than 10° of latitude south of the subtropical convergence and well to the north of the Polar Front, although a number of the species extend farther south either in the central-eastern Pacific or Chilean sectors; a group of species with northern limits of distribution generally within the subtropical convergence region and with southern limits of distribution generally within the Polar Front region although one species, Lampanyctus achirus, regularly extends south of the Polar Front; a group of species with northern and southern limits of distribution generally within about 5° of latitude north and south of the Polar Front, although juveniles of the same species regularly extend farther north; and, finally, a group of species with northern limits of distribution, particularly of adults, generally near the Polar Front and with southern limits of distribution extending, in at least the Pacific and Indian sectors, far south of the Polar Front to near the Antarctic continent. Thus within a broad spectrum of individual distribution a number of common patterns can be recognized. These patterns are correlated with prominent hydrologic features such as the subtropical convergence and Antarctic Polar Front. Five major patterns and several subpatterns are here delimited. These are diagrammatically presented in Figure 56. The species whose distributions conform to one or another of the proposed patterns are listed in Appendix 2.

Pattern 1

Thirty-nine species have southern limits of distribution generally at or just north of the region of the subtropical convergence. Included under pattern 1 are several species not represented in the study material but which have been reported from south of 30°S by Bussing [1965], Becker [1967a], Nafpaktitis and Nafpaktitis [1969], and Craddock and Mead [1970]. These species are discussed in the previous section.

Several trends emerge from a consideration of latitudinal and longitudinal ranges of the species involved in pattern 1. On the basis of latitudinal distribution, these species may be more or less artificially differentiated into those normally found no farther north than 20°-30°S (subpatterns A and B) and into those with ranges that regularly extend as far north as the equator (subpatterns C and D). In terms of longitudinal distribution these same species may be subdivided into those with a restricted longitudinal range (i.e., in part of one ocean; subpatterns A and C) and those found in more than one ocean (subpatterns B and C) and those found in more than one ocean (subpatterns B and D). In addition, a number of species cannot be categorized because of taxonomic problems of fragmentary distributional data.

Subpattern A. Protomyctophum chilensis, P. sp. D, Symbolophorus sp. D, and Diaphus danae are latitudinally and longitudinally restricted to the southwestern and southeastern Pacific, respectively. The decision to include P. sp. D and S. sp. D in this pattern rather than in pattern 2 was influenced by the fact that they have primarily been collected with pattern 1 species in areas where pattern 2 species are absent.

Subpattern B. Hygophum hygomi, Myctophum phengodes, Gonichthys barnesi, Lampandena dea, L. chavesi, L. speculigera, Diaphus parri, D. effulgens, Lampanyctus pusillus, and Scopelopsis multipunctatus have restricted latitudinal distributions in the southern hemisphere, but, in contrast with subpattern A species, they have relatively broad longitudinal ranges. Diaphus parri, M. phengodes, L. dea, S. multipunctatus, and G. barnesi are restricted to the southern hemisphere. The rest of the species in this subpattern are represented by populations in the North Atlantic Ocean.

Subpattern C. Lepidophanes guentheri, Lampanyctus parvicauda, and Triphoturus mexicanus have broad latitudinal but narrow longitudinal ranges. Lepidophanes guentheri is broadly distributed in the tropical-subtropical Atlantic Ocean and occurs south of 30°S in the western Atlantic. Lampanyctus parvicauda and T. mexicanus are restricted to the eastern Pacific.

Subpattern D. Electrona rissoi, Benthosema suborbitale, Diogenichthys atlanticus, Notolychnus valdiviae, Taaningichthys bathyphilus, Lampanyctus alatus, Ceratoscopelus warmingi, Lobianchia dofleini, and Notoscopelus resplendens are characterized by broad longitudinal and latitudinal distributions. Lobianchia dofleini, N. resplendens, and D. atlanticus are known from 40°N to 40°S on the periphery of subtropical gyres in the Atlantic Ocean and from south of 20°-30°S in the Indian and Pacific oceans. Notoscopelus resplendens and D. atlanticus also occur as far north as 35°N in the eastern Pacific Ocean. Populations of E. rissoi are known from 40°N to 40°S in the eastern Atlantic and eastern Pacific oceans, from 30°-40°S in the Indian and southwestern Pacific oceans, and from the equatorial Indian and northwestern Pacific oceans. The five remaining subpattern D species are characterized by even wider distributions, and unlike the four aforementioned forms, they are found in most latitudes of the Indian Ocean. Except for N. valdiviae, however, each is apparently absent from at least one major region of the tropical-subtropical World Ocean.

In addition to the species grouped by subpat-

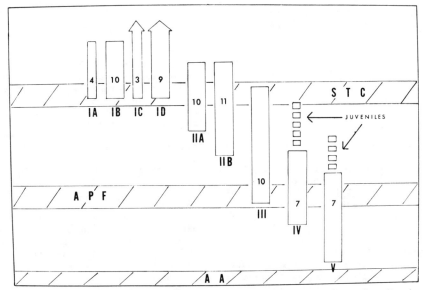

Fig. 56. Diagrammatic summary of patterns of lanternfish distribution south of 30°S. Arabic numerals refer to the number of species assigned to each pattern of distribution. See text.

terns, there are a number of pattern 1 species, that, because of taxonomic problems and fragmentary data, cannot be assigned to any one subpattern with any degree of certainty. These species include Hygophum macrochir group, Myctophum nitidulum, Symbolophorus spp., Centrobranchus nigroocelatus, Bolinichthys supralateralis, B. indicus, Lampanyctus sp. B, L. sp. C, L. niger-ater group, Lobianchia gemellari, Diaphus sp. B, D. mollis, and D. termophilus. Diaphus termophilus, D. mollis, L. sp. B, M. nitidulum, C. nigroocelatus, and L. gemellari are each known from one locality south of 30°S.

Pattern 2.

This pattern is shown by 21 species distributed in and near the subtropical convergence. None of these species commonly occur as far south as the Polar Front and, with the following exceptions, north of 30°-35°S: Symbolophorus sp. B, Lampanyctus australis, Metelectrona ventralis, and Diaphus ostenfeldi are known from isolated localities near 20°S in the Benguela Current; Lampichthys procerus and Diaphus sp. A from about 20°S in the Peru Current; and Lampanyctus lepidolychnus from approximately 25°S, 60°E. The distributions of Hygophum brunni and Lampanyctus iselinoides, both restricted to the waters off Chile between 30° and 50°, are also included in pattern 2.

Although data are admittedly sparse between 85° and 135°W, it appears that two subpatterns can be defined on the basis of the presence or absence of the species in that area.

Subpattern A. Protomyctophum subparallelum, Metelectrona sp. A, Hygophum hanseni, Lampanyctus sp. A, L. sp. D, L. intricarius, Diaphus ostenfeldi, D. sp. A, Hintonia candens, and Loweina interrupta apparently have continuous distributions extending from the Atlantic sector to the eastern Pacific sector. Lampanyctus macdonaldi, which was included in pattern 3 has a distribution

quite similar to several subpattern A species, and additional collections in subantarctic waters of the Atlantic and Indian sectors may warrant its placement in this subpattern. Like L. macdonaldi, three subpattern A species, L. intricarius, L. interrupta, and P. subparallelum are represented by closely related populations in the North Atlantic Ocean. A fourth species, D. sp. A, is represented by a closely related population in the North Pacific Ocean.

Subpattern B. Electrona paucirastra, Metelectrona ventralis, Symbolophorus sp. B, Symbolophorus sp. C, Hygophum brunni, Lampadena notialis, Lampanyctus australis, L. lepidolychnus, Lampanyctus iselinoides, Lampichthys procerus, and Lampanyctodes hectoris are apparently not continuously distributed from the Atlantic sector to the eastern Pacific sector. In fact, four of these species seem to occur on or near continental shelves: L. iselinoides and H. brunni are limited to the waters off Chile, S. sp. C to waters off eastern New Zealand, and L. hectoris to the near-shore waters of Africa, Australia, New Zealand, and Chile. The rest of the species are known from the Atlantic, Indian, and Pacific (western part) sectors. Five of the latter are known also from the eastern Pacific off Chile. Lampichthys procerus may eventually prove to have a distribution more appropriately assigned to subpattern A.

Pattern 3

This pattern is shown by the distributions of Protomyctophum normani, P. sp. C, P. parallelum, Electrona subaspera, Symbolophorus sp. A, G. piabilis, G. sp. A, G. sp. C, Lampanyctus achirus, and L. macdonaldi. These species are distributed primarily between the Polar Front and the subtropical convergence with L. achirus occurring also south of the Polar Front in the Australian, Pacific, and Atlantic (western part) sectors, as

well as north of the subtropical convergence off
Peru. Only three or four pattern 3 species are
found in subantarctic waters throughout the
'Southern Ocean': P. parallelum, E. subaspera,
L. achirus, and possibly L. macdonaldi. The rest
of the species are either restricted to or excluded from part of the Pacific sector.

Pattern 4

This pattern is shown in the distributions of
Protomyctophum tenisoni, P. sp. A, P. sp. B, P.
andriashevi, Electrona carlsbergi, Gymnoscopelus
bolini, and G. fraseri. These species are more
or less restricted to the region of the Antarctic
Polar Front, although juveniles of some may be
found as far north as the region of the subtropical convergence. Their southern limits of distribution generally extend only a short distance
south of the Polar Front. Near the eastern extension of the Weddell-Scotia confluence, however,
adults of six of these species extend as much as

3°-10° of latitude south of the Polar Front; P.
sp. B is the exception.

Pattern 5

This pattern is shown in the distribution of
Protomyctophum anderssoni, P. bolini, Electrona
antarctica, Gymnoscopelus braueri, G. opisthopterus, G. aphya, and G. sp. B. These species are
more or less restricted to Antarctic waters south
of, and in the region of, the Antarctic Polar
Front, although juveniles of some species extend
well north of the Polar Front into subantartic
waters. All seven species are known from well to
the south of the Polar Front in Pacific and Australian sectors, whereas only three, E. antarctica, G. braueri, and G. opisthopterus, are also
known from far south of the Polar Front in most
of the Atlantic and Indian sectors. This pattern
includes all but one of the myctophids commonly
collected south of the region of the Polar Front.
Since Lampanyctus achirus ranges as far north as
the subtropical convergence, it has been included in pattern 3.

A comprehensive biogeographic study should be based on the least subjective analysis of data derived from the least limiting sampling program. Limitations inherent in the data presented in the previous chapter permit only a preliminary analysis of the distribution of lanternfishes south of 30°S. Four obvious limitations, common to most biogeographic studies of oceanic organisms are the following.

1. Inadequate understanding of basic hydrological processes, such as those that occur in the vicinity of the subtropical convergence, imposes one restriction on the analysis. It was emphasized in a previous chapter that although the northern limits of the west wind drift may be conceptually distinct, they remain physically indistinct and arbitrarily defined with a region of transition present between the west wind drift and the subtropical gyres. This region was approximated by reference to the 34.6 and 34.8/mil isohalines at 200 m. These isopleths certainly represent only a portion of the transition region, most probably its southern limits, and even so only to a variable degree, depending on longitude. For example, the apparent distribution of these isohalines relative to distribution limits of patterns 1, 2, and 3 species differs in various sectors of the study area. This is particularly obvious in the central-eastern Pacific, where subpattern 2A species and a majority of pattern 3 and subpattern 1B and 1D species show a definite southward shift in relation to the somewhat arbitrary position of the transition region. A clearer understanding of oceanographic processes in the region of the subtropical convergence would facilitate a more precise definition of the transitional region. For the present, however, I must assume that the simplest explanation for much of the longitudinal variation in distribution is due to the differences in hydrology and an inadequate definition of hydrologic boundaries such as the transition region rather than ecological differences between conspecific populations of lanternfishes.

Other areas which appear to be of particular zoogeographic importance and which require additional oceanographic study are the Antarctic Polar Front, the Falkland Current, the subsurface circulation in the subantarctic region, and the Pacific subantarctic waters. The apparent existence of a generally unrecognized westward-flowing current in the latter region exemplifies our inadequate understanding of subantarctic water movement.

2. The geographic distribution of samples imposes the most potentially serious restrictions on an analysis such as the present one. Relatively large areas south of 30°S have not been sampled by large midwater trawls, and some of them have only been superficially sampled by other, less efficient devices. The present data were obtained from a variety of collecting gear at various roughly determined, or undetermined, depths over an extended period of time. It is known that population densities of many pelagic

species are correlated with water mass changes and that many species show marked temporal and spatial variation in vertical and horizontal distribution. At present, the quantitative analysis necessary for a more accurate interpretation of myctophid distribution south of 30°S would be very difficult if at all feasible. The availability of comparable data from the entire study area and appropriate computerized procedures would make such a quantitative analysis possible. The present study analyzes the apparent overall distribution of lanternfish species as determined by the presence and absence of capture data within and outside a particular geographic area. In almost all instances, the absence of species outside their apparent area of distribution is reinforced by the presence of different species with different patterns of distribution.

3. Taxonomically confused and uncertain cases impose additional restrictions on this analysis. The major reason for difficulties at the species level is inaccessibility of myctophids for observation and experimentation and the consequent necessity of defining species solely on the basis of morphology. The specific status and relationships of allopatric populations of morphologically similar species are difficult to ascertain. In addition, only recently has enough material been accumulated for the time-consuming studies necessary for interpretation of phylogenetic relationships of myctophids at the supraspecific level. Much more work remains to be done.

4. The lack of general biological information on lanternfishes, particularly regarding their reproductive biology, imposes a subtle, yet serious, limitation on this study. Pelagic organisms are usually classified as either planktonic or nektonic, depending on their ability to swim. Although it is conceptually useful, in reality this dichotomy is frequently inadequate. Are lanternfishes planktonic or nektonic? Different students have different opinions, but most consider them a special category of the plankton or the nekton, e.g., micronekton [Blackburn, 1968] or macroplankton [Parin, 1970]. Myctophids vary in size from less than 30 mm (Notolychnus valdiviae) to more than 250 mm (Gymnoscopelus bolini) and undoubtedly also vary in swimming ability. All species, however, are strictly planktonic, i.e., incapable of active sustained horizontal migration during their larval and postlarval stages, and I know of no evidence which suggests that any species of lanternfishes undergoes an active horizontal migration at any time in its life cycle. Becker [1967a, b] and others have pointed out the similarity between the distribution of lanternfishes and the distribution of invertebrate zooplankton. Ekman [1953] suggests that differences between the planktonic and nektonic modes of life impose functional differences in distribution. The same author also remarks that the area of distribution of a nektonic species includes a breeding area, a foraging area, and sometimes a separate area where planktonic young are carried. According to Parin [1970],

the area of distribution of a planktonic species may include a basic distribution range, a nonsterile region of settlement, and a sterile region of settlement. He further defines the basic distribution range as a 'more or less closed area, where a species may subsist for a long time irrespective of its presence or absence in other parts of the ocean.' The nonsterile region of settlement is the 'entire part of a breeding area that extends beyond the basic range...' This occurs when 'an area of the water that is congenial for reproduction constantly flows out of the circulation that serves as the basic range but does not return to it.' Sterile regions of settlement are areas 'where animals of dependent populations can exist but cannot breed...' and as such 'do not strictly belong to the range of the species.' Nafpaktitis [1968] and O'Day and Nafpaktitis [1965] have documented and discussed the expatriation to sterile regions of several myctophid species in the North Atlantic Ocean.

It is imperative to recognize that limits of distribution are based on peripheral records for individual species, although in some cases, peripheral records are considered so anomalous that they are ignored and that the relative densities and reproductive fates of the peripheral populations are in most cases unknown. The possibility that peripheral populations might occur at low densities or as expatriates could account for the somewhat gradual transition between the patterns proposed in the last chapter. Despite numerous limitations and unknowns, however, the fact remains that myctophid species south of 30°S reveal patterns of distribution which correlate with hydrologically recognizable regions and which thus define several biogeographic complexes.

The Complexes

The study area includes the southern periphery of the warm water circulation of the great southern anticyclonic gyres, as well as the east and west wind drifts. The major hydrologic regions, defined and discussed previously, are zonal and include, from north to south, (1) a warm water region, (2) the transitional region of the subtropical convergence, (3) the subantarctic region, (4) the region of the Antarctic Polar Front, and (5) the Antarctic region. These five regions correspond to distributional patterns 1, 2, 3, 4, and 5, respectively.

The Warm Water Species

The 39 species whose distributions show pattern 1 represent a significant component of the warm water myctophid fauna of the World Ocean. The single feature which unites all of the species in pattern 1 and distinguishes them from the 90-100 other warm water myctophid species is their known occurrence south of 30°S.

The warm water fauna is taxonomically well differentiated from the fauna of the Southern Ocean, i.e., the species whose distributions have been assigned to patterns 2-5. The region of the subtropical convergence marks the southern limits of distribution for 14 of the 21 genera in pattern 1. Eleven of these, which are relatively small genera (1-5 species), including Centrobranchus,

Gonichthys, Diogenichthys, Benthosema, Notolychnus, Triphoturus, Taaningichthys, Ceratoscopelus, Lepidophanes, Scopelopsis, and Notoscopelus, are each represented by one species in the study area. Three, including Myctophum (15 species), Bolinichthys (5 species), and Lobianchia (2 species), are each represented by two species. The remaining 22 species in pattern 1 belong to seven speciose genera (7-50 species) that are also represented in the more southern patterns. Most of these warm water forms, however, are not closely related to their southern congeners. Electrona rissoi is very distinct from its four congeners, all of which are restricted to the southern patterns. Protomyctophum chilensis and P. sp. D are more closely allied with populations in the North Pacific Ocean and with each other than with the numerous southern species of Protomyctophum. Both pattern 1 Hygophum species, H. hygomi and H. macrochir group, are closely allied with other warm water forms and not with the southern H. brunni or H. hanseni. The two or three species of Symbolophorus in pattern 1 are most closely related to each other, although all six or seven species represented in the study area, together with two Atlantic species, form a natural taxonomic unit within the genus. Lampadena chavesi is distinct from all of its congeners, whereas the genus Lampadena chavesi is distinct from all of its congeners, whereas L. dea and L. speculigera are allied with L. notialis, a pattern 2 species. Of six species of Diaphus in pattern 1, only D. parri is closely related to a more southern species; it is related to D. sp. A, a pattern 2 species, and the North Pacific species D. theta. Finally, only two of the six Lampanyctus species in pattern 1 seem to be closely allied with southern forms; L. alatus and L. sp. C are related to L. australis and L. sp. D, respectively, of pattern 2.

To review, at least 39 species in 14 genera of warm water myctophids show a southern limit of distribution in the region of the subtropical convergence. Only 7 genera of lanternfishes are represented by species in pattern 1 and by different species in the more southern patterns. Only 5 of the 22 pattern 1 species in these 7 genera seem to be closely related to their more southern congeners, and in all five instances, the relatives are species assigned to pattern 2, which is limited to the region of the subtropical convergence. With the exception of these 5 species, the warm water fauna is remarkably distinct from that assigned to patterns 2-5. Further collections made in southern subtropical waters may disclose the presence of some pattern 2 species, particularly rare species such as Loweina interrupta, and warrant their reassignment to pattern 1.

A number of the warm water biogeographic patterns which exist are undoubtedly represented by distributions of the various lanternfishes assigned to pattern 1. The extensive biogeographic literature on warm water pelagic organisms, including myctophids, however, is pervaded with conflicting concepts and terminology and is much in need of critical review. This task is beyond the scope of the present study, and the following discussions will merely highlight general biogeographic features of the warm water pelagial by reference to selected studies. The reader is

referred to Knox [1970], McGowan [1971], Beklemishev [1967], and Parin [1970] for more detailed
discussion.

Specific areas or centers of distribution of
many pelagic taxa are frequently consistent with
water mass patterns and boundaries [McGowan,
1971]. Perhaps the most basic generalization
that can be drawn on a world wide scale is the
existence of cold water biotas separated by a
warm water biota, within which similar zonal patterns of distribution occur in the northern and
southern hemispheres. The periphery of the warm
water biota is associated with polar edges of
subtropical gyres. This has been or can be shown
for the epipelagic ichthyonekton [Parin, 1970],
euphausiids [Brinton, 1962; Mauchline and Fisher,
1969], myctophids [Becker, 1964b], tropical deep-
sea animals [Ebeling, 1967], and the plankton of
the Pacific Ocean [Beklemishev, 1967]. Parin
[1970] points out that although his proposed distribution scheme for the epipelagic ichthyonekton
is broadly similar to the scheme proposed by Beklemishev [1967] for the shallow plankton of the
Pacific Ocean, zonal patterns within the warm
water fauna are much less marked in the former.
Parin recognizes only two basic patterns of distribution in the warm water epipelagic ichthyonekton: one which extends across the equator
between the approximate position of the summer
surface 20°C isotherms of both hemispheres and a
second which comprises the areas between the summer and winter surface 20°C isotherms in the
northern and southern hemispheres, respectively.
He also emphasizes that the warm water ichthyonekton is remarkably uniform throughout the Atlantic, Indian, and Pacific oceans. In contrast,
many midwater fish species of such families as
the Melamphaidae and Myctophidae have restricted
warm water distribution [Ebeling, 1962, 1967;
Becker, 1967a, b]. Becker [1967b] compares patterns of lanternfish distribution in the Pacific
Ocean with Beklemishev's [1967] patterns and finds
them to be generally the same. These patterns,
as well as similar patterns proposed by other
workers for the euphausiids and other planktonic
groups [Brinton, 1962; Mauchline and Fisher, 1969;
McGowan, 1971; Ebeling, 1962, 1967] correlate
with watermass patterns and boundaries which are
primarily, but not panoceanically, zonal in warm
waters of the World Ocean. The basic patterns of
distribution are primarily influenced by the central and equatorial water masses, the two basic
types of water which occur between subarctic and
subantarctic latitudes. They comprise a mosaic
which includes (1) species limited to equatorial
waters, frequently to relatively limited and hy-
drographically unique waters, e.g., the eastern
part of the equatorial Pacific Ocean, (2) species
rather broadly distributed in central waters, but
which, for the most part, are excluded from equatorial waters, particularly in the Indian and
eastern Pacific oceans, (3) species with a variety
of relatively restricted distributions in central
waters, including forms limited to or concentrated
in the relatively infertile mid parts of the
central water masses of the Atlantic, Indian, and
Pacific oceans, forms limited to the equatorial
periphery of the central water masses, and forms
limited to the polar periphery of the central
water masses, (4) species with broad distributions
extending throughout all or most of the equator-

ial and central waters, (5) neritic or nearshore
species with limited distributions frequently
similar to the distributions of littoral species,
and (6) species with distributions intermediate
between or a combination of the aforementioned
patterns or distributions. In addition, some
authors recognize a transitional pattern of distribution associated with the boundary region
between central and subpolar water masses. This
pattern will be discussed in the section on transitional water species.

Except for Triphoturus mexicanus, Lampanyctus
parvicauda, and Protomyctophum chilensis, which
are found south 30°S only off Chile, most pattern
1 species are members of warm water faunas associated to varying degrees with central waters.
Interestingly, except for the three forms mentioned above, most of these species do not occur
in the waters off Chile, where central water are
held well offshore by longitudinal currents
carrying equatorial and subantarctic waters.

The distributions of the seven species in subpatterns 1A and 1C correspond quite closely to
discrete water masses which adjoin the Southern
Ocean. Triphoturus mexicanus and L. parvicauda,
both of which are restricted to the eastern Pacific, occur as far south as the area of transition
between equatorial and modified subantarctic
waters off Chile. The antitropical P. chilensis
is restricted to this area of transition. Symbolophorus sp. D, Promyctophum sp. D, and possibly Lampanyctus sp. C seem to be restricted to
the eastern South Pacific central water mass,
whereas Diaphus danae appears to be limited to
the western South Pacific central water mass.
Lepidophanes guentheri, which is distributed
throughout North and South Atlantic central water
masses, occurs south of 30°S in the region of the
Brazil Current.

Subpattern 1B includes ten species which appear
to be limited to a relatively narrow band of central waters adjacent to the subtropical convergence (subtropical waters), whereas the nine species assigned to pattern 1D are also known from
equatorial latitudes in at least part of their
respective ranges. My attempt at defining these
two subpatterns should not obscure the fact that
an almost gradual transition exists from species
with a southern subtropical distribution to species with a broader tropical-subtropical range.
Subpattern 1B includes five species limited to
the southern hemisphere and five species with
disjunct populations in subtropical waters
(25°-45°N) of the North Atlantic Ocean (bisubtropical). Perhaps significantly, Lampadena
speculigera and Lampanyctus pusillus of the latter group are known from isolated records as far
north as 15°S in the southeastern Atlantic,
whereas Myctophum phengodes and Scopelopsis multipunctatus of the stricly southern group are
known from as far north as 10°-15°S in the southeastern Pacific and Indian oceans. Diaphus parri,
D. effulgens, Hygophum hygomi, Lampadena speculigera, and possibly M. phengodes appear to be absent from subtropical waters of the southwestern
Atlantic Ocean.

Three subpattern 1D species have distributions
similar to the bisubtropical subpattern 1B species
in the North Atlantic Ocean and throughout most
of the southern hemisphere. They differ from the
subpattern 1B species, however, in that they are

also known to occur in some equatorial latitudes: _Lobianchia dofleini_, _Notoscopelus resplendens_, and _Diogenichthys atlanticus_ are known from approximately 40°N to 40°S in the eastern Atlantic Ocean. _Lobianchia dofleini_ is not known from north of 25°S in the remainder of the World Ocean. _Notoscopelus resplendens_ and _D. atlanticus_ are also known from a broad latitudinal band in the eastern Pacific Ocean, between 40°N and 40°S, although, except for larval stages, both species are apparently excluded from the Pacific equatorial water mass. Significantly, all three species occupy wide latitudinal ranges only in the eastern Atlantic Ocean where subtropical waters converge at the equator and a discrete equatorial water mass does not exist.

A fourth subpattern 1D species, _Electrona rissoi_, appears to be primarily distributed near the eastern, equatorial, and parts of the polar peripheries of all subtropical gyres. It may be excluded from the relatively sterile core of the central water masses.

The five remaining subpattern 1D species are distinguished by their known occurrence in all latitudes of the Indian Ocean. Of these, however, only _Notolychnus valdiviae_ is known from virtually all of the warm waters of the World Ocean. _Lampanyctus alatus_ appears to be absent from the Pacific Ocean east of 165°E, from the southwestern Atlantic Ocean, and from the central core of the North Atlantic subtropical gyre. _Benthosema suborbitale_ seems to be excluded from all but the equatorial edges of the central waters of the Pacific Ocean east of 165°W and, except near Africa, is probably absent from the South Atlantic Ocean. _Taaningichthys bathyphilus_ is broadly distributed but may be absent from the South Atlantic Ocean. _Ceratoscopelus warmingi_, although apparently excluded from the range of its close relative, _C. townsendi_, in the eastern North Pacific Ocean, is probably a ubiquitous warmwater species. A majority of subpattern 1B and 1D species, i.e., 12 of 19, are known from the region of the Agulhas Current.

To summarize, pattern 1, which represents more than a fourth of all warmwater lanternfish species, includes a strong element of zonally distributed southern subtropical species, the ranges of which tend to merge with those of more broadly distributed species. Most of these species are absent from the near-shore waters of Chile and many appear to be excluded from the southwestern Atlantic Ocean.

The Transitional Water Species

Numerous planktonic organisms have been found to be limited to or concentrated in the boundary regions between central and subpolar waters [Brinton, 1962; Johnson and Brinton, 1963; Mauchline and Fisher, 1969; Gibbs, 1968; Frost and Fleminger, 1970; McGowan, 1971]. Brinton [1962] has suggested that these regions, along with areas of mixing between modified subpolar and equatorial waters in the Peru Current and California Current, be termed transitional biogeographical zones in the southern hemisphere and in the North Pacific Ocean. This proposal has been accepted by Johnson and Brinton [1963], Mauchline and Fisher [1966], and McGowan [1971]. Other workers have rejected or only partly agreed with this concept of tran-

sitional biogeographical zones. Beklemishev [1965] recognizes transitional zones in both hemispheres of the Pacific Ocean, but only that of the northern hemisphere coincides with Brinton's. Beklemishev's southern zone occurs in subantarctic waters. Brodskey [1965] and Knox [1970] argue against the recognition of transitional biogeographical zones, the latter author preferring to call them warm temperate zones. Knox, however, questionably assumes that the component species are distributed primarily in subtropical waters. Although some species considered to be transitional by Brinton and others may be primarily limited to subtropical waters, the majority would appear to be limited to colder transitional waters. Brinton has actually included subtropical euphausiids in a central water assemblage. Although the term transitional may be somewhat misleading, there are several factors which favor its acceptance. As argued by Brinton on the basis of euphausiid distribution and as substantiated by lanternfish distributions [Becker, 1967b, for the North Pacific], distinct complexes of species occur in the transitional areas between central and subpolar waters and between modified subpolar modified subpolar and equatorial waters off western America, in the North Pacific Ocean and the southern hemisphere. These complexes are distinct from complexes which are restricted to adjacent subtropical waters, e.g., subpattern 1B of lanternfishes, and adjacent subantarctic waters, e.g., pattern 3 of lanternfishes. This fauna occurs at the boundary between major patterns of circulation of the World Ocean, and as such the connotation of the term transitional is meaningful. The lack of a more descriptive term in the literature, the added confusion that would arise from the introduction of a new term, and the use of the term transitional in a considerable body of literature over the last decade lead me to its use for the 21 species assigned to pattern 2.

Twelve genera are represented by the 21 species assigned to pattern 2. The predominantly warm water genera _Lampadena_, _Hygophum_, _Diaphus_, and _Loweina_ contribute a total of six species, their southernmost representatives to this pattern. The monotypic genera _Lampichthys_, _Lampanyctodes_, and _Hintonia_ and the genus _Metelectrona_ (two species) are known only from the region of the subtropical convergence. _Protomyctophum_, _Electrona_, _Symbolophorous_, and _Lampanyctus_, which contribute one, one, two, and six species, respectively, to pattern 2, are known from more northern and southern patterns.

Transitional water species include several groups with divergent geographic affinities. Four species are represented by very similar populations in the northern hemisphere, primarily to the north of northern populations of bisubtropical warm water species: _Protomyctophum subparallelum_, _Loweina interrupta_, and _Lampanyctus intricarius_ in the North Atlantic Ocean and _Diaphus_ sp. A in the North Pacific Ocean. Four transitional water species, including _Lampanyctus australis_, _L._ sp. D, _Lampadena notialis_, and one of the aforementioned forms, _Diaphus_ sp. A, are closely related to pattern 1 species. Only two transitional water species, _Lampanyctus_ sp. A and _Electrona paucirastra_, are most closely related to more southern species which are assigned to pattern 3. The transitional water pattern also includes four species pairs. Hygophum hanseni and H. brunni are closely related to each other and relatively distinct from their warm water congeners. Metelectrona ventralis and M. sp. A are also very closely related. Symbolophorus sp. B and S. sp. C, also most closely related to each other, are closely allied with other congeners in patterns 1 and 3 and in the Atlantic Ocean north of the study area. The fourth species pair comprises Lampanyctus lepidolychnus and the antitropical L. intricarius. The relatively distinct transitional water species L. iselinoides may also be related to L. lepidolychnus. The four remaining transitional water species include Lampichthys procerus, Lampanyctodes hectoris, and Hintonia candens, which are obviously well differentiated from other myctophid species, and Diaphus ostenfeldi, which is quite distinct from its numerous warm water congeners. It is apparent that a diverse assemblage of lanternfish species is restricted to the transitional region of the subtropical convergence in the southern hemisphere. The assemblage comprises a striking endemic fauna, about one third of which is closely allied with adjacent faunas and about one fifth of which is also known from similar latitudes of the northern hemisphere.

Several interesting geographic trends are reflected in the distributions of transitional water species. Surprisingly, few species are circumglobal. Subpattern 2B includes four species with localized distributions as well as seven species with much wider distributions, all of which, however, appear to be excluded from much of the Pacific sector, particularly in the central part. Hygophum brunni and Lampanyctus iselinoides are endemic in the area between 30° to 50°S off Chile, where modified subantarctic waters flow northward over southward-flowing equatorial waters. Symbolophorus sp. C is endemic to regions where northward and southward currents converge off the east coast of New Zealand. Lampanyctodes hectoris is known from transitional waters near the coasts of Chile, New Zealand, Australia, and Africa. The seven remaining subpattern 2B species, with the exception of Electrona paucirastra, generally occur throughout the transitional zone in the Atlantic, Indian, and western Pacific sectors. Electrona paucirastra appears to be excluded from the western part of the Atlantic sector. The absence of these seven species throughout most of the central Pacific sector and the presence of populations of five of them near the Chilean coast is particularly intriguing: Electrona paucirastra, Metelectrona ventralis, Symbolophorus sp. B, Lampanyctus australis, and Lampichthys procerus are also known from the area in which H. brunni and L. iselinoides are endemic, whereas Lampadena notialis and Lampanyctus lepidolychnus are not known from east of 160°E and 165°W, respectively. In contrast, many subpattern 1A species which do occur throughout the central Pacific sector are apparently absent from the western part of the Atlantic sector and from the near-shore waters of Chile. Protomyctophum subparallelum, Metelectrona sp. A, Lampanyctus sp. A, L. intricarius, L. sp. D, Hintonia candens, and Loweina interrupta are not known from the western part of the Atlantic sector, and M. sp. A, Hygophum hanseni, L. sp. A, L. sp. D, and L. interrupta are excluded from Chilean waters. In addition, only three pattern 2 species, L. australis, L. lepidolychnus, and H.

hanseni, are presently known from the Agulhas portion of the transitional region near Africa. Diaphus ostenfeldi and D. sp. A seem to be the only transitional water species with circumglobal distributions. And, as was mentioned previously, Lampanycthys procerus may eventually prove to be circumglobal. Interestingly, not only are subpattern 2A species continuously distributed across the central Pacific sector, but also the majority are found farther south here, some as far as the Polar Front, than in other sectors. In this area, transitional and subantarctic waters flow to the east and submerge, mixing with a westward flow of subantarctic water from near the subtropical convergence. Although the vertical distribution of various transitional water species is not accurately known, subpattern 2A species generally appear to be deeper living than subpattern 2B species, the majority of the latter having been collected at the surface.

The Subantarctic Species

The ten species whose distributions correspond to pattern 3 are almost exclusively found in subantarctic waters. Two species of Lampanyctus are exceptions: L. achirus is commonly found south of the Polar Front and off Peru and L. macdonaldi is also known from the North Atlantic Ocean. It was mentioned previously that the distribution of the latter species is somewhat intermediate between pattern 3 and subpattern 2A and that additional data from the Atlantic and Indian sectors may warrant its placement with the transitional water species. Many other pelagic taxa are known to include numerous species similarly restricted to subantarctic waters [McGowan, 1971], and a variety of biogeographic terms have been proposed to describe the subantarctic region [Knox, 1970], including subantarctic, notalian, south cold temperate, austral, and transitional. I will refer to members of pattern 3 as a subantarctic species because this seems to be the least ambiguous and most descriptive term available.

Five genera, including both subgenera of Protomyctophum, are represented by subantarctic species. Protomyctophum (Hierops), Symbolophorus, and Lampanyctus are represented by their one, one, and two southernmost species, respectively. Protomyctophum (Protomyctophum) and Gymnoscopelus are represented by their two and three northernmost species, respectively. Electrona, which is known from farther north and farther south, is represented by one subantarctic species.

Subantarctic species form several groups with divergent geographic affinities. Two species pairs occur in subantarctic waters: Protomyctophum normani and Gymnoscopelus piabilis are most closely related to P. sp. C and G. sp. C respectively, and both species pairs are less closely allied with pattern 4 species, P. sp. B and G. sp. B. Gymnoscopelus sp. A, is most closely related to the pattern 4 species G. fraseri. Protomyctophum parallelum, Symbolophorus sp. A, and possibly the antitropical Lampanyctus macdonaldi are relatively distinct from, but allied with, species groups distributed in or near transitional areas of both hemispheres. Electrona subaspera and L. achirus are also most closely related to the transitional water species E. paucirastra and L. sp. A, respectively.

The most striking feature of the distributions assigned to the subantarctic pattern is the distinctive character of the Pacific sector. Here Symbolophorus sp. A and Gymnoscopelus sp. A are endemic to the large region of counterclockwise water movement from which four additional subantarctic species are generally excluded. Two of the latter species, P. sp. C and G. piabilis, are each represented by disjunct populations east and west of the area occupied by S. sp. A and G. sp. A. A third species, P. normani, does not occur east of 120°W, whereas G. sp. C is limited to near the Polar Front east of 135°W. Only three or four species, P. parallelum, E. subaspera, L. achirus and possibly L. macdonaldi, are found throughout the subantarctic region. Interestingly, the largest size groups of these species are found farther south than smaller size groups, and in E. subaspera, younger stages are found farther north than older stages.

Inspection of the distributions of P. parallelum, S. sp. A, G. sp. C, G. sp. A, E. subaspera, L. macdonaldi, and juvenile L. achirus between 75°W and 120°W reveals an additional interesting feature. In relation to the Polar Front, their southern limits of distribution are several degrees of latitude farther north between 75° and 90°W, and in the case of the latter five species even north of the Polar Front, than between 90° and 120°W, where several are found south of the Polar Front.

The Antarctic and Antarctic Polar Front Species

The 14 species whose distributions make up patterns 4 and 5 are closely associated with waters in the region of and south of the Polar Front, respectively. Similar distributions have been indicated for many pelagic taxa, including euphausiids [Mauchline and Fisher, 1969], chaetognaths [David, 1965], salps [Foxton, 1965], fishes [Andriashev, 1962; Parin, 1970], and other groups. Such distributions are generally referred to as Antarctic distributions. Relatively few data from the Antarctic and Polar Front regions of the eastern Atlantic and Indian sectors, as well as considerable taxonomic similarity and distributional congruence, prompt me to include both patterns in a single biogeograhic unit, the Antarctic/Antarctic polar front complex. The juveniles of these species are characteristically distributed farther north than, and not as far south as, adults.

Only three myctophid genera are represented in the Antarctic/Antarctic polar front complex: Protomyctophum (Protomyctophum) by six species, Electrona by two species, Gymnoscopelus (Gymnoscopelus) by four species, and Gymnoscopelus (Nasolychnus) by two species. Gymnoscopelus is endemic to Antarctic waters, G. (Nasolychnus) and P. (Protomyctophum) to Antarctic and subantarctic waters, and Electrona, discounting the markedly distinct warm water species E. rissoi, to Antarctic, subantarctic, and transitional waters. With the exception of the subantarctic species L. achirus, which is known to occur also in Antarctic waters, the Antarctic region is inhabited by exclusively Antarctic and Polar Front species of genera or subgenera which are essentially endemic to the Southern Ocean.

Three species pairs are included in the Antarctic/Antarctic polar front complex. Gymnoscopelus braueri, G. aphya, and Protomyctophum tenisoni are closely related to G. opisthopterus, G. bolini, and P. sp. A, respectively. Five species, P. sp. B, P. andriashevi, G. sp. A, G. sp. B, and E. antarctica are closely related to subantarctic species. The three remaining species, P. andersonni, P. bolini, and E. carlsbergi are relatively distinct from their congeners, although E. carlsbergi may be allied with E. antarctica.

Pattern 4, the Antarctic Polar Front species, includes Protomyctophum tenisoni, P. sp. A, P. sp. B, P. andriashevi, Electrona carlsbergi, Gymnoscopelus bolini, and G. fraseri. Larger individuals of these species appear generally to be restricted to the region of the Polar Front, whereas juveniles are generally known from farther north. There is a northward shift in the distributions of many of these species in relation to the Polar Front between 75° and 90°W. The southern limits of P. tenisoni, P. sp. A, P. sp. B, E. carlsbergi, and G. fraseri and the northern limits of larger size groups of P. tenisoni, P. andriashevi, G. fraseri, and G. bolini are several degrees of latitude farther north in this area than in adjacent areas. In addition, there seems to be general southward shift in distribution near the eastern edge of the Weddell-Scotia confluence, where, except for P. sp. B, the southern limits of larger size groups of Antarctic Polar Front species extend a variable but considerable distance south of the Polar Front. Interestingly, the distributions of most of these species appear to be interrupted in the hydrologically complex area between 45° and 60°W. Here, there is an apparent counterclockwise movement of water which extends northwestward from the northern tongue of the Weddell-Scotia confluence around South Georgia to the Polar Front and perhaps beyond. One population of P. sp. A is largely restricted to this area and occurs from the Argentine shelf through the region near South Georgia to south of the Weddell-Scotia confluence. The remaining pattern 4 species appear to be absent from, or relatively uncommon, in all or part of this area.

Antarctic region species in pattern 5, unlike pattern 4 species, are known to be distributed broadly in Antarctic waters in at least the Pacific sector of the Southern Ocean. Only Electrona antarctica, Gymnoscopelus braueri, and G. opisthopterus are distributed in the southernmost areas sampled, and even these are excluded from the Ross Sea, where the pelagic icthyofauna consists primarily of a single species of the predominantly benthic family Nototheniida [Dewitt, 1970]. Gymnoscopelus opisthopterus, unlike E. antarctica and G. braueri which are distributed nearly throughout the sampled Antarctic region, has not been collected in northern Antarctic waters of the Atlantic and Indian sectors east of 30°W. Here it is apparently restricted to waters south of 55°-69°S. In contrast, the four remaining Antarctic region species, Protomyctophum bolini, P. anderssoni, Gymnoscopelus aphya, and G. sp. B, which are broadly distributed in Antarctic waters of the Australian and Pacific sectors, except near 150°W at the northern edge of the Ross Sea Gyre, have been collected only near the Polar Front, north of the area of occurrence of

G. opisthopterus in the Atlantic and Indian sectors east of 30°W. In the western part of the Atlantic sector, P. bolini and G. aphya extend a short distance south of the Weddell-Scotia confluence, and P. anderssoni extends as far south as the Weddell-Scotia confluence. This confluence also marks the southern limits of younger stages of G. braueri, G. aphya, G. opisthopterus, P. bolini, and E. antarctica, as well as all growth stages of the subantarctic species L. achirus. Gymnoscopelus sp. B, like Salpa gerlachei, a pelagic tunicate [Foxton, 1965] does not appear to extend through the Drake Passage into the western part of the Atlantic sector. Unlike the tunicate, however, G. sp. B is known from the eastern part of the Atlantic and Indian sectors. The northern limits of the adult stages of most pattern 5 species are generally near the Polar Front, although they range somewhat farther north between 70° and 90°W and south of Australia. As is mentioned above, G. opisthopterus is exceptional in that its northern limits of distribution are considerably south of the Polar Front in the eastern Atlantic and Indian sectors. The southern limits of younger stages are farther north in relation to the Polar Front between 75° and 90°W than in other areas.

Maintenance of the Patterns
of Distribution

The distribution of any population is determined primarily by several interdependent factors: capacity for dispersal and movement, tolerance for ambient physical and biological environmental parameters, and, of course, evolutionary history. The assortment of numerous lanternfish populations into relatively discrete patterns of distribution associated with heterogenous hydrographic regions suggests that the influence of oceanic circulation on their distribution is fundamental. This is particularly evident after a summary comparison of distributional limits within and between the basic patterns discussed above. Such a comparison reveals a mosaic of responses by different species to the same hydrologic factors.

Thus many warm water species, seven of which are included in this study (subpatterns 1A and 1C), are restricted to discrete hydrographic regions adjoining the Southern Ocean from which a varying number of more broadly distributed warm water species are excluded. Two eastern equatorial Pacific species occur as far south as central Chile, and the southern population of an antitropical species is limited to near-shore waters of central Chile where relatively distinct equatorial water flows poleward beneath a northward flow of subantarctic water; the majority of warm water species are excluded from this area, as are central waters with which they are associated. One warm water species found in central waters of the Atlantic Ocean ranges south to about 40°S with the Brazil Current in the southwestern Atlantic Ocean; the majority of subtropical warm water species are absent from this area where waters are more tropical than subtropical in nature. The eastern and western South Pacific central water masses also harbor endemic species and exclude a limited number of broadly distributed species. Other relationships of hydrology and distributions of warm water species have been

described in the literature and reviewed above, and additional relationships will undoubtedly be uncovered with further collection and analysis. One such relationship suggested by the present analysis that is worth mentioning here is evident in equatorial latitudes of the eastern Atlantic Ocean where waters flowing equatorward from subtropical latitudes of both hemispheres converge and a distinct equatorial water mass is lacking. The fauna there includes a strong complement of species associated with subtropical waters, three of which are otherwise restricted to subtropical latitudes and two of which are the endemic Lampadena pontifex and an undescribed species of Symbolophorus, both members of species groups otherwise restricted to subtropical or higher latitude waters.

A number of the features that characterize the distributions of warm water species are complemented by transitional water species. Subpattern 2B species, for example, are generally excluded from the transitional region adjacent to the eastern South Pacific central water mass. Five species are represented by disjunct populations in the subantarctic equatorial transitional region off south central Chile, primarily south of the ranges of warm water populations and more broadly distributed populations in the Atlantic, Indian, and western Pacific sectors; two species are endemic to the same region near Chile; and one species is endemic to an area of convergence and mixing off the east coast of New Zealand. The majority of transitional water species in subpattern 2A, which do occur in the eastern-central Pacific sector, also contrast with subpattern 2B species by their absence from the near-shore waters of Chile and, apparently, from the western part of the Atlantic sector near the Brazil Current. Many of these species are carried southeastward to their southernmost limits of distribution with transitional components of the counterclockwise movement of water in the Pacific subantarctic region.

Four subantarctic species are excluded from all or most of the counterclockwise-moving subantarctic water in the east-central Pacific sector, an area in which two species are endemic. Two species are represented by disjunct populations off southern Chile, primarily to the south of transitional water species and more broadly distributed populations in and west of the western part of the Pacific sector, one species is absent from the Pacific sector east of 130°W, and the fourth species is restricted to southernmost subantarctic waters east of 135°W by southeastward-moving transitional and subantarctic waters. The latter current almost certainly accounts for the unusually high latitude that other subantarctic species in this same region are found. In contrast, the southern limits of most subantarctic species are several degrees of latitude farther north between 75° and 90°W than in adjacent areas of southward flow of water in the subantarctic gyre and in the Cape Horn Current. At the same time the northern and/or southern limits of Antarctic Polar Front species (pattern 4) as well as the northern limits of older stages and southern limits of younger stages of Antarctic region species (Pattern 5) are generally farther north between 75° and 90°W than in other longitudes. These features may be attributable to what seems to be a strong flow of Antarctic surface water across the Polar Front, as is evidenced by the relatively cold isotherms which course northeastward from the Polar Front region between 75° and 90°W [Gordon and Goldberg, 1970].

The distributions of Antarctic Polar Front species appear to be primarily affected by the dynamics of the Polar Front. Significantly, the complex region of the Polar Front between 45° and 60°W that receives a mass of water from the region of the Weddell-Scotia confluence apparently harbors an endemic population of one Polar Front species to the exclusion of remaining species. More broadly distributed Antarctic species are influenced to varying degrees by movements of water in and from the east wind drift, particularly from the Ross and Weddell seas, where most of these species are restricted to northern antarctic latitudes.

It is evident, then, that lanternfishes are similar to passively drifting organisms such as the euphausiids, whose distributions receive identity from systems of circulation [Brinton, 1962]. By making the reasonable assupmtion that efficient reproduction by a planktonic genotype is constrained within an inherited tolerance for ambient environmental parameters such as light, temperature, food, etc., beyond which decreased reproduction, sterility, or death results, the intrinsic role played by planktonic populations in maintaining distributions within discrete systems of circulation may be considered. A planktonic organism spawned into a moderate current of 5 cm/s (about 0.1 kn) would be carried more than 1500 km during its first year of life. Although this would not impose problems for populations resident in circular patterns of water movement such as those that maintain the central water masses, much of the ocean is characterized by nongyral currents capable of carrying resident populations at particular depths across major hydrographic boundaries into alien environments. The currents at different depths in these areas, however, move at different velocities and frequently in different directions, and the vertical distribution of any planktonic population would undoubtedly influence its horizontal distribution. Indeed, in many areas which harbor endemic populations the currents at different depths flow in opposing directions, e.g., the Southern Ocean, equatorial latitudes, and many neritic areas. Possibly significant in this context is the fact that one of the most ubiquitous characteristics of oceanic midwater organisms is vertical migration, the frequency of which may be diel, ontogenetic, or seasonal. One of the various theories proposed to explain the significance of diel vertical migration, all of which may be true to a limited degree [Mauchline and Fisher, 1969], suggests that diel migration probably evolved as a means of giving plaktonic forms a degree of independence from the environment, enabling individual organisms to continually sample new environments [Hardy, 1956]. It is also possible that, in addition to such a function on a microgeographic scale, diel migration could increase the probability that genotypes would be exposed to the same environment as their forebearers. Populations adapted to vertical patterns of circulation could use them to maintain themselves in discrete macrogeographic areas to which they are adpated.

Mackintosh [1937] and others have pointed out that ontogenetic and seasonal vertical migrations can substantially explain the limited geographic distribution of a number of midwater organisms. Similarly, the apparent seasonal and, for that matter, diel and ontogenetic migrations of Electrona antarctica can substantially account for its remarkable fidelity to Antarctic waters; while in the upper layers, individuals would be transported northward with surface waters, in the lower layers, they would be carried southward with returning Antarctic intermediate water or deep water. Other Antarctic species appear to undergo ontogenetic vertical migrations which would substantially account for their horizontal distributions. In most species, younger stages are found in shallower depths and to the north of older stages, although in some instances juveniles may be deeper. Junveniles would be carried northward with surface or intermediate waters and by subsequently altering depth would be carried southward by the subsurface water flow. Similar cases may also be made for some subantarctic species. In addition, it is significant in this respect that the numerous restricted populations present in the study area occur where adjacent vertical and/or horizontal currents flow in opposite directions. Off Chile, for example, restricted populations of warm water species, transitional water species, and subantarctic species replace each other from north to south in the region where modified subantarctic water flows over southward-flowing equatorial waters. It would be fruitful to investigate and compare the vertical distributions of different lanternfish species within such an area as well as assemblages from different areas. For example, species broadly distributed in central waters would not necessarily undergo vertical migrations for the purpose of maintaining their distributions, and this may be reflected in their patterns of vertical distribution. An investigation of the relationship between vertical and horizontal patterns of distribution, however, must wait until more precise data are available. For the present, it can be assumed that the horizontal distributions of lanternfish populations are a result of interrelationships between an inherited tolerance for biotic and abiotic environmental parameters, patterns of vertical distribution, and patterns of oceanic circulation. Evolution would proceed according to the availability of habitats, i.e., patterns of the circulation capable of conserving passively drifting populations [Brinton, 1962]. Fractions of individual populations at least large enough to maintain themselves would simply not be carried by currents transgressing their particular habitats. Different species would undoubtedly have different patterns of vertical distribution as well as different limits of tolerance for environmental parameters. This conceptual framework provides a substantial rationale for the distributional features described above, including the transition noted between the major patterns of distribution, as well as such features as expatriation, changes in density, anomalous records, or absence of any species beyond a particular limit.

Evolution of the Complexes

The following discussion is largely summarized by McGinnis [1977]. Adult and larval morphology indicate that the Myctophidae are probably derived from an ancestor that also gave rise to the family Chlorophthalmidae [Paxton, 1972; Moser and Ahlstrom, 1970]. Fossil myctophids are known from numerous Tertiary and Quaternary deposits as far back as the Eocene era. They are apparently absent from Cretaceous strata [Goody, 1969]. The transition from the Mesozoic to the Cenozoic era, which marked a dramatic change in terrestrial and neritic biotas, seems to have been accompanied by similar change in oceanic midwater biotas, as is evidenced by the evolution and radiation of the Myctophidae family.

Fraser-Brunner [1949] suggests that the family probably originated in the Antarctic region, primarily because the assumed primitive genera Protomyctophum (sensu stricto) and Electrona are nearly restricted to the Southern Ocean. The candidacy of these genera for the most primitive and generalized forms in the family has recently been challenged [Moser and Ahlstrom, 1970, 1972, 1974]. Bolin [1939], on the other hand, has suggested a warm water origin. I would agree with Andriashev [1962], who supports Bolin's hypothesis but adds that the Tethys Sea undoubtedly played an important role in the early stages of lanternfish evolution. The invasion of oceanic midwaters may or may not have proceeded from the neritic zone by way of an intermediate epipelagic phase [Parin, 1970]. In any case, its close relationship with the benthic Chlorophthalmidae would indicate that the Myctophidae family is probably derived from a benthic ancestor. Chlorophthalmids are characterized by planktonic larvae [Mead, 1966] that could be viewed as having the form of a generalized lanternfish larva [Moser and Ahlstrom, 1970, p. 142]. An ancestral population on which natural selection would press an extended larval phenotype toward neoteny is conceivable. Such a population would be preadapted to discrete patterns of oceanic circulation. The primary biological significance or ecological advantage of a pelagic larval stage to a benthic population is believed to be dispersal [Mayr, 1963, p. 607; Mileikovsky, 1971, p. 200]. The in situ evolution of an extended larval stage, however, would almost certainly be due to selective pressure for increasing the probability of delivery of a prepared phenotype to a suitable, and presumably, ancestral habitat. Whether the ancestral habitat was a vast area such as the Tethys Sea, which nurtured a relatively homogenous warm water fauna during the Cretaceous and early Tertiary era or some more restricted area, the ancestral population would have been evolving in relationship with discrete patterns of oceanic circulation prior to the elimination of the benthic adult state; the population would have been preadapted for a pelagic adaptive zone when different aspects of the niche forced the pelagic stage to take over reproduction. In any case, Fitch's [1969] analysis of Tertiary lanternfishes indicates that populations attributable to at least one recent genus were present in Eocene seas and that the family had achieved much of its present level of generic differentiation by Miocene.

Orders, families, and even genera of most pelagic organisms are characteristically distributed in more than one major region of the World Ocean, and consequently, the pelagial is very weakly differentiated into faunistic regions [Ekman,

1953; McGowan, 1971]. A consideration of the distribution of lanternfish genera and relatively distinct subgeneric groups (of related species), however, does reveal a limited generic differentiation of lanternfishes in various regions of the World Ocean. Becker [1964b] has proposed two major lanternfish complexes in the World Ocean, a temperate-cold water complex including faunas beyond the subtropical gyres in the northern and southern hemispheres and a tropical-warm water complex associated with the anticyclonic gyres. My analysis of the distribution of lanternfish species south of 30°S, which has revealed the existence of basic patterns of distribution associated with Antarctic, transitional, and warm waters allows a more precise description of lanternfish complexes in the southern hemisphere, which can be compared to other regions. This analysis supports the basic dichtomy into cold and warm water faunas proposed by Becker. For convenience, Protomyctophum (Hierops) and P. (Protomyctophum) are treated as separate genera in the following discussion.

The most diverse assemblage of lanternfish genera occurs in central and equatorial waters of the World Ocean (Table 2). This assemblage includes 15 genera which are not shared, and 7 genera which are shared, with the subantarctic and/or transitional water assemblages in the southern hemisphere. Only one lanternfish genus is represented in both the warm water pattern and in the antarctic pattern, and in this case, the single species, Electrona rissoi, is very distinct from its numerous congeners, all of which are restricted to the transitional, subantarctic, or Antarctic patterns.

Dioenichthys, Gonichthys, Centrobranchus, Notolychnus, Taaningichthys, Lepidophanes, Bolinichthys, Triphoturus, Parvilux, Lobianchia, Scopelopsis, Benthosema, Notoscopelus, Ceratoscopelus, and Myctophum are not represented in the more southern patterns. Ten of these genera are also excluded from similar regions north of the anticyclonic gyres in the northern hemisphere, whereas Benthosema, Notoscopelus, and Ceratoscopelus are each known to have one species endemic to the boreal waters of the North Atlantic Ocean and Myctophum is known to have one species which ranges in the area. One species of Parvilux is restricted to transitional waters of the NE Pacific Ocean.

Hygophum, Lampadena, Loweina, and Diaphus are predominantly warm water genera which also have species in the transitional region. Protomyctophum (Hierops), Symbolophorus, and Lampanyctus have species with transitional and with subantarctic distributions. Several of these genera are represented by species in transitional and/or subpolar waters of the northern hemisphere. Most of these genera contain groups of related warm water species which are well differentiated from their cold water congeners. Becker [1965] and Krefft [1970] have reported species groups of Hygophum and Lampadena, respectively, that are endemic to warm waters. Similar warm water species groups undoubtedly exist in Symbolophorus, Lampanyctus, and Diaphus. The warm water species of P. (Hierops) are more closely related to each other than to their cold water congeners, and Loweina, as was mentioned previously, may eventually prove to a be a warm water genus.

Table 2. Number of Lanternfish Species by Genus in Major Regions of the World Ocean

Genus	No. of Species	Antarctic-Antarctic Polar Front	Sub-antarctic	Southern Subtropical Convergence	Warm Water	Cold North Atlantic	Cold North Pacific
Gymnoscopelus	9	6	3				
Protomyctophum	8	6	2				
Electrona	5	2	1	1	1		
Metelectrona	2			2			
Hintonia	1			1			
Lampanyctodes	1			1			
Lampichthys	1			1			
Loweina	4?			1	4		
Lampadena	8			1	7		
Hygophum	9			2	7		
Lampanyctus	40		2	6	many	X	X
Symbolophorus	14?		1	2	6-8	1	1
Hierops	8		1	1	3	1	2
Diaphus	50			2	many	X	X
Myctophum	15?				12-14	1	
Ceratoscopelus	3				2	1	
Notoscopelus	5				4	1	
Benthosema	6?				4-5	1	
Scopelopsis	1				1		
Lobianchia	2-3				2-3		
Triphoturus	3-4				3-4		
Bolinichthys	5				5		
Lepidophanes	2				2		
Taaningichthys	3				3		
Notolychnus	1				1		
Centrobranchus	4				4		
Gonichthys	4				4		
Diogenichthys	3				3		
Parvilux	2				1		1
Tarletonbeania	1						1
Stenobrachius	2						2
Dorsadena	1						1

The numbers of species not from the 'Southern Ocean' are approximate. X denotes undetermined number of species.

Most myctophid genera occurring in warm waters are widely distributed in the Atlantic, Indian, and Pacific oceans. Notable exceptions are Scopelopsis, which is restricted to southern subtropical waters; Triphoturus, which is more or less restricted to equatorial waters; P. (Hierops), the warm water species of which are restricted to Pacific subtropical or subpolar equatorial transitional waters; Parvilux, the warm water species of which is restricted to the eastern Pacific Ocean; and Lepidohanes, which is restricted to the Atlantic Ocean. In addition, however, there are undoubtedly species groups of other genera which are restricted to discrete areas, particularly to central or equatorial waters. Individual warm water species appear to have relatively limited distributions, Notolychnus valdiviae and possibly Ceratoscopelus warmingi being the only species presently known to be more or less ubiquitous in the anticyclonic gyres of the Atlantic, Pacific, and Indian oceans.

A number of myctophid genera are cold water forms; i.e., they are essentially excluded from warm equatorial and central waters. Tarletonbeania and Stenobrachius are endemic to subarctic and transitional waters of the North Pacific Ocean, whereas Dorsadena may be restricted to North Pacific transitional waters. A number of species of Symbolophorus, Diaphus, Lampanyctus, and P. (Hierops) are also endemic to this area. Of these, Diaphus theta is very similar to and P. (Hierops) thompsoni is most closely related to but distinct from transitional water populations in the Southern Ocean.

The monotypic genera Lampichthys, Lampanyctodes, and Hintonia and the genus Metelectrona as well as 16 species of other genera are endemic to the transitional region of the Southern Ocean. Fifteen of the latter species are members of 7 genera that also occur in warm waters. Most of these species, however, are relatively distinct from their warm water congeners. Several pairs of species are allied with each other, and four species are actually most closely related to or are indistinguishable from populations restricted to higher latitudes of the northern hemispheres: P. (Hierops) subparallelum to the North Atlantic P. (Hierops) articum and to a lesser degree, the North Pacific P. (Hierops) thompsoni; (Diaphus sp. A to the North Pacific D. theta; Lampanyctus intricarius which is also known from the North Atlantic Ocean; and Loweina interrupta, which may actually be bisubtropical in the Atlantic Ocean. The transitional region also harbors one species of Electrona, a genus which occurs only in Antarctic and subantarctic regions, discounting the divergent warm water species E. rissoi.

A third cold water generic assemblage occurs south of the transitional region in the southern hemisphere. The numerous species of Gymnoscopelus and Protomyctophum are endemic to Antarctic or subantarctic waters. Also found in these waters are three species of Electrona and the four southernmost, all subantarctic, species of Lampanyctus, P. (Hierops), and Symbolophorous. One of the latter species, L. macdonaldi, also occurs in the cold waters of the North Atlantic Ocean.

The only cold water region lacking endemic genera is the boreal North Atlantic Ocean. The fauna of this region includes a number of endemic species, some of which are apparently derived from the warm water Myctophum, Benthosema, Notoscopelus, and Ceratoscopelus, and some of which are very closely related to or are conspecific with forms in the Southern Ocean, i.e., P.(Hierops) arcticum, Symbolophorus veranyi, Lampanyctus intricarius, and L. macdonaldi.

Thus the warm water region and the following cold water regions each harbor endemic lanternfish genera: the North Pacific Ocean, the southern transitional region, and the Antarctic-subantarctic regions combined (Table 2). The assemblages of which these genera components show varying degrees of relationship to each other. The transitional water and Antarctic-subantarctic assemblages share one genus, Electrona. The warm water and transitional water assemblages share three genera, Hygophum, Lampadena, and Loweina. Four genera, Diaphus, Lampanyctus, Symbolophorus, P. (Hierops), are shared between the warm water, transitional water, and North Pacific assemblages, three of which are also represented in the subantarctic assemblage. Cold water species of this latter group are generally more closely related to each other than to their warm water congeners.

Returning now to the fossil record, Fitch [1969] reported eight myctophid genera to have been well established in North American seas south of 40°N throughout the Miocene era: Diogenichthys, Hygophum, Lampanyctus, Lepidophanes, Myctophum, Notoscopelus, Symbolophorus, and Diaphus, the latter genus extending back in the fossil record to the Eocene era of California. These genera are presently restricted to or show their greatest diversity in warm waters. Fitch further reported six of the preceding genera and seven additional genera from the Pliocene era of California. The additional genera are the primarily warm water Benthosema, Ceratoscopelus, Lampadena, and Electrona rissoi, the essentially equatorial Triphoturus, the boreal North Pacific Stenobrachius and Tarletonbeania, and the subpolar transitional subtropical P. (Hierops). At the specific level the Pliocene fauna is remarkably modern. Twelve of the 13 species reported by Fitch are subarctic, transitional, central, or equatorial forms presently found off California. Danil'chenko [1967] reported the appearance and extinction of the lanternfish genus Eomyctophum during the middle Eocene and middle Oligocene eras of Europe, repectively. In the genus he

includes three Oligocene species which show marked osteological differences from each other. Recent genera are known from the later Tertiary era of Europe [Danil'chenko, 1967]. The New Zealand Tertiary era is rich in myctophid material (J. E. Fitch, personal communication 1971), and might eventually yield evidence on the evolution of myctophids in the southern hemisphere. For the present, however, a perspective of myctophid evolution must primarily be based on the abundant fossil record of the northeastern Pacific Ocean and its relationship to paleogeography and paleoceanography.

The oceanography and geography of the World Ocean during the early Tertiary era undoubtedly differed from that of the present time. An exchange of water between the Atlantic and Indian basins via the Tethys Sea may have continued from the Cretaceous era into the Oligocene era [Fell, 1967]. The Panamanian seaway between the Atlantic and Pacific basins remained open until the Pliocene era [Lloyd, 1963], allowing an exchange of equatorial waters, and an injection of water from the Atlantic Ocean may have stmiluated a more vigorous and broad equatorial circulation in the eastern Pacific than now exists [Hays et al., 1969]. In contrast with equatorial latitudes, an exchange between ocean basins via a circumpolar current in the Southern Ocean may not have begun until the Tertiary era. Agreement, however, is lacking on when the major continents and ocean basins approached their present positions. Fell [1967] indicated that by mid-Jurassic, world geography differed from present geography primarily in the greater extent of epicontinental seas but not in the relative positions of continents and ocean basins. Frakes and Kemp [1972], however, suggest that South America probably had contact with Antarctica until the Tertiary era, and the same is possibly true for Australia [Watkins and Kennett, 1971]. Either connection would result in a significantly different hydrology within the Southern Ocean. Frakes and Kemp [1972] have reconstructed early Tertiary climates and geography from a variety of data. They concluded that a circumpolar current did not exist until the early Oligocene era, when the relatively warm conditions that had characterized high latitudes during the Eocene era were altered to conditions more like the present. Cooling of the more broadly tropical World Ocean of the Eocene era to a warmer than present minimum in the Oligocene era and a subsequent warming to a maximum in the Miocene era is evidenced by trends of Tertiary faunas in widely scattered localities, including the northeastern Pacific Ocean [Addicott, 1970], New Zealand [Devereaux, 1967], and the Southern Ocean [Margolis and Kennett, 1970]. The cooling which commenced in the Moicene era culminated in the glaciations of the Pliocene and Pleistocene eras, whereas a major east Antarctica ice sheet accumulated in the Miocene era [Shackleton and Kennett, 1975]. Post Miocene world climate is reviewed and related to paleoclimatic trends in Antarctic deep-sea cores by Hays [1968], who cites evidence of two periods of rapid cooling in Antarctic waters about 3.0-2.5 and 0.7 m.y. ago. The latter cooling appears to be synchronous with cooling in the North Pacific, equatorial Pacific, and North Atlantic oceans. Between these two periods, near the Plio-Pleistocene boundary there is evidence of the initiation of large-scale upwelling in Antarctic waters, without a corresponding change in temperature. During the last 350,000 years the Southern Ocean has undergone a number of less marked fluctuations in temperature and concurrent latitudinal displacements of the Antarctic Polar Front and ice shelf. The most recent southward retreat of the Polar Front to its present position may have covered more than 5° of latitude [Kennett, 1970; Be, 1969]. Kennett [1970] suggests eight intervals of warming of subantarctic waters during the last 1.2-1.3 m.y.; one interval at 0.5-1.04 m.y. ago which may have been significantly warmer than the present may be related to an unusually warm interval in the North Pacific Ocean [Kent et al., 1971]. Fewer fluctuations of temperature reflected in Antarctic sediments than in equatorial Pacific sediments may indicate local oceanographic events [Hays, 1968].

On the basis of the preceding discussion a number of inferences can be made.

1. The Myctophidae family had achieved considerable differentiation as early as the Eocene era and had approached its present level of generic differentiation by the Miocene era. Although Eocene and Oligocene fossils need comprehensive study and revision, the apparent existence in the Eocene era of one species of the predominantly Oligocene genus Eomyctophum and two species attributable to Diaphus, as well as four additional species listed from the Eocene era of Europe [see Fitch, 1969, p. 17], would indicate that considerable divergence had already occurred within the family. The remarkably high number of recent species of Diaphus in warm waters of the World Ocean that are undoubtedly most similar to Eocene seas would seem to corroborate its early evolution. The three species attributed to Eomyctophum by Danil'chenko [1967] show substantial, probably generic, osteological differences, e.g., jaw length relative to orbit diameter and vertebral counts, that had already evolved by the middle Oligocene era. The presence of at least eight recent warm water genera in the warm seas that bathed western North America during the Miocene indicates that by that time the family approximated its present degree of differentiation.

2. Oceanographic conditions conducive to the evolution of precursors of regionally restricted genera in the present World Ocean had been in existence as early as the Oligocene era. The establishment of a circumpolar Southern Ocean and a period of climate approaching that of the present would have allowed the evolution of nuclear populations within the warm waters of the still connected Atlantic, Pacific, and Indian basins and within the cold waters of the southern hemisphere, the North Pacific Ocean, and possibly the North Atlantic Ocean. In regards to the North Atlantic, however, it seems possible that its relatively small size and warm temperatures during the Oligocene era [Frakes and Kemp, 1972] may have inhibited the development of an incipient cold water fauna. Eventual study of European fossils, including Oligocene forms presently ascribed to Eomyctophum and Miocene forms ascribed to extant genera may indicate whether a cold water fauna differentiated in the North Atlantic during the Paleogene era. Although evidence is lacking, southern cold water faunas may conceivably have

developed from two Eocene faunas previously adapted to high-latitude circulations of the then isolated or semiisolated southern Pacific and Atlantic-Indian sectors, respectively. According to the reconstructions of Frakes and Kemp [1972], the Pacific sector was considerably colder than the Atlantic-Indian sector during the Eocene era and also during the Oligocene, and it seems possible that endemic transitional water and Antarctic-subantarctic genera may have been derived from Atlantic-Indian and Pacific populations, respectively. A broad and vigorous equatorial circulation that may have existed in the eastern Pacific Ocean during the Miocene era and, presumably, the Oligocene era, would have effectively isolated the cold water faunas of the southern hemisphere and the North Pacific Ocean. The boreal shallow water molluscan fauna of the North Pacific had become largely provincial by the Oligocene era, and a greater part of the North Pacific may have become available to its embryonic boreal fauna at this time [Briggs, 1970]. The absence of the boreal North Pacific genera Stenobrachius and Tartletonbeania from the California Miocene era is almost certainly due to the warm water conditions under which the California deposits were laid down rather than their absence from cold North Pacific waters.

Eventual analysis of myctophid fossils from the New Zealand Tertiary and Pleistocene eras may provide direct information on the evolution of Tertiary myctophid faunas of the Southern Ocean. As was noted above, paleoclimatic interpretations indicate a similarity between New Zealand, the Southern Ocean, and the Northeastern Pacific in the Tertiary era. The fact that New Zealand molluscan faunas were wholly subtropical during the Miocene era and that they received a strong influx of cold water forms during the Pliocene era indicates that the subtropical convergence was south of New Zealand during the Miocene era [Knox, 1963]. This decreases the probability of finding Miocene fossil of Antarctic-subantarctic genera. It seems more probable that the subtropical genus Scopelopsis, and, possibly, the transitional water Lampichthys, Hintonia, Lampanyctodes, Metelectrona, and P. (Hierops) will be found in the New Zealand Miocene era. These transitional water genera will also probably be found in deep-water Pliocene localities. In any case, a high degree of endemism and diversity indicate that myctophid assemblages in cold waters of the southern hemisphere have developed from two incipient faunas over a considerable period of time, one in association with the transitional waters and the other in association with the Southern Ocean proper. Their suggested early evolution is corroborated by the apparent antiquity of the neritic Antarctic fish fauna [Norman, 1938] and by the evidence of an austral marine fauna as early as the Mesozoic era [Fleming, 1962]. Within the transitional water assemblage, the genera Metelectrona, Lampanyctodes, Lampichthys, and Hintonia represent an old element, and as was suggested above, they may represent relicts of a fauna that could have evolved in high latitudes of the Atlantic-Indian sector during the Eocene era. The remaining transitional water species, which show varying degrees of relationship with the warm water fauna, or in the case of Electrona, with the Antarctic and subantarctic faunas, the majority of which

appear to be members of subgeneric groups of species that have speciated in transitional and, in
some case, adjacent waters, represent younger
elements. Within the Antarctic and subantarctic
assemblages, Protomyctophum, Gymnoscopelus, and
Electrona represent an old element, possibly derived from an Eocene fauna that could have evolved
in high latitudes of the Pacific sector during
the Eocene era. This group has undergone a remarkable radiation in the Southern Ocean, and is
presently represented by 20 species in Antarctic
and subantarctic waters. A total of 4 species of
Lampanyctus, Symbolophorus, P. (Hierops), which
are found in subantarctic waters, make up the
remaining and more recent elements of this assemblage.

3. The climatic changes of the later Tertiary
and Pleistocene eras were accompanied by an overall cooling with episodic fluctuations in temperature and hydrology and the emergence of the middle American isthmus that would have markedly
affected the distributions of populations adapted
to previous patterns of oceanic circulation and
temperature. This is indicated by lanternfishes
of diverse geographic affinity that appear in the
California Pliocene era, particularly the boreal
North Pacific Stenobrachius and Tarletonbeania,
the equatorial Triphoturus, and the antitropical
P. (Hierops). Fitch [1969] found Stenobrachius
in ten Pliocene and Pleistocene localities, usually in conjunction with Tarletonbeania, which
was present in seven of the same deposits. Triphoturus (three localities) and P. (Hierops) (four
localities) were always found with both previous
genera but were both present at the same locality
only once. The earliest record of P. (Hierops)
is from a Pliocene deposit which, according to
Fitch, was laid down when ocean temperatures were
much colder than temperatures at the same latitude
today. I suggest that as the climate deteriorated
and the California Current became more intense
and/or cooler, nuclear boreal species would have
been carried further south. During a particularly
cold episode, P. (Hierops) transgressed the equatorial region from one hemisphere to the other,
most likely from south to north. As the Panamanian seaway closed, the present character of the
eastern tropical Pacific became established, and
undercurrents of equatorial water moving poleward
formed transitional regions with subpolar waters
and carried equatorial forms as far as California.

The distributions of southern lanternfish populations were undoubtedly also altered by fluctuations in temperature and in the position of
hydrographic regions such as the Polar Front during the Neogene era. Elucidation of the evolutionary processes which accompanied such alterations, molding Tertiary faunas into existing lanternfish complexes, will require additional study
of the vertical as well as the horizontal distribution of lanternfish species, particularly of
larvae and juveniles, and clarification of the
systematic relationships of both recent and fossil species. Such studies will necessarily be
approached at a subgeneric level. It is apparent
from the present analysis, however, that many
features of lanternfish distribution can be interpreted in the context of an allopatric model
of speciation in which physical or spatial al-
teration of a pattern of oceanic circulation in-

habited by a planktonic population can result in
disjunct populations on which differential selection can operate. Subsequent alterations can
bring about partial or total sympatry. Brinton
[1962] has argued for the evolution of existing
patterns of distribution in euphausiids, particularly central and transitional species, within
such a context; he suggests that paleoclimatic
changes have been acompanied by the alternate
coalescence at and retraction from equatorial
latitudes of central and transitional waters,
which allow the central or transitional faunas of
each hemisphere access to and subsequent isolation
in the other hemisphere. The antitropical affinities of a considerable number of lanternfishes,
which are restricted to or overlap the transitional region of the subtropical convergence, can
be explained by Brinton's hypothesis of equatorial
confluence of northern and southern transitional
zones during the past. A particularly strong
affinity between South and North Atlantic populations indicate that the most recent interchange
may have been more recent and/or colder in the
Atlantic Ocean than in the Pacific Ocean. Antitropical species include one subantarctic as well
as several transitional and subtropical ones,
each of which is morphologically indistinguishable
from or very similar to its northern hemisphere
sibling. Antitropical forms include one transitional species which occurs at an unusually low
latitude in the Peru Current system and two additional forms which are restricted to Pacific
subtropical waters and equatorial-subantarctic
transitional waters, respectively. An older, and
perhaps colder, interchange of faunas in the
Pacific Ocean may be evidenced by the relative
distinctness of another transitional species, P.
(Hierops) subparallelum, from its North Pacific
counterpart, P. (Hierops) thompsoni, and the appearance of P. (Hierops) in a particularly cold
Pliocene fossil locality; P. (Hierops) subparallelum is less easily distinguished from P.
(Hierops) arcticum, its counterpart in the North
Atlantic Ocean. Briggs [1966, 1970] has summarized additional evidence for a relatively more
dramatic and recent cooling of the Atlantic Ocean,
and McIntyre [1967] has found evidence of extensive southward displacement of boreal Atlantic
coccolithophorids during the Pleistocene era.

The numerous disjunct and regional distributions of transitional water, subantarctic, and,
to a lesser degree, Antarctic species indicate
that an allopatric mechanism of speciation may
also operate strictly within the Southern Ocean.
This is particularly evident in the Pacific sector
where numerous transitional water and subantarctic
species are represented by populations off the
Chilean coast and near New Zealand but are absent
from the intervening area. Such distributions
can also be explained by downstream dispersal of
larvae, but it would seem that the hydrology of
the Pacific subantarctic region, if I have interpreted it correctly, would tend to impede such a
dispersal of transitional water and subantarctic
species. Antarctic species, which are particularly instructive, are the closely related P.
(Protomyctophum) tenisoni and P. sp. A (Figures
7, 8) Protomyctophum sp. A appears to be composed
of two morphologically distinct and allopatric
populations, although this remains to be substantiated. One population is found within a unique

hydrographic region near South Georgia, and the other population, for which the data are few, appears to be distributed with P. tenisoni in the region of the Antarctic Polar Front. The most striking difference between the two populations is found in the morphology of the male supracaudal luminous gland. Males in the South Georgia population have glands very similar to those of P. tenisoni, whereas those of the population sympatric with P. tenisoni are markedly different. I assume that the sexually dimorphic caudal glands function in mating behavior and would suggest that the unique morphology of the males sympatric with P. tenisoni represent an example of the evolution of a premating isolating mechanism. If this is true, the two populations should probably be regarded as specifically distinct. Morphological divergence of secondary luminous tissue is common in genera found in the study area, including P. (Protomyctophum), Electrona, Hygophum, Symbolophorus, and Gymnoscopelus. Such divergence occurs in a number of related species which have largely allopatric but contiguous distributions as well as in largely sympatric species. Fleminger [1967] has observed evidence of reinforcement in largely allopatric but contiguously distributed copepods and a lack of reinforcement in related but completely allopatric species. He

reasons that these features can be interpreted in the context of past fluctuations in oceanography and that the contiguous species were probably more broadly sympatric during particularly intense climatic episodes at which time natural selection favored the evolution of reproductive barriers. It can be noted at this time that patterns of morphological divergence and distribution in some lanternfish populations south of 30°S seem to represent varying stages of this process and thus corroborate Fleminger's hypothesis. Such features, in addition to the evidence for fluctutations in hydrology and temperature, e.g., latitudinal movements of the Polar Front, lead me to suspect that such fluctuations are largely responsible for the present day nature of lanternfish assemblages in the Southern Ocean. Protomyctophum sp. A and P. tenisoni may represent a case in which subsequent to isolation and differentiation a change in oceanography has led to limited introduction of one species into the range of other species and the evolution of even a third population. It is hoped that future studies on the ecology, systematics, and distribution of lanternfishes will facilitate an interpretation of the evolutionary processes which have molded the present distribution and composition of lanterfish complexes south of 30°S.

Summary

Lanternfishes (family Myctophidae) represent an unparalleled evolutionary radiation of primary carnivores in the midwaters of the World Ocean. They are abundant, widespread, and speciose, and, in addition, lanternfishes are common in post Mesozoic marine fossil deposits. Knowledge of the distribution of lanternfish taxa should provide insight into the evolution of midwater communities. Extensive collections by the research vessels USNS Eltanin and HMS Discovery have provided numerous data for analysis of the composition and distribution of the lanternfish fauna in the World Ocean south of 30°S.

The World Ocean south of 30°S includes the southern periphery of the southern subtropical gyres of the Atlantic, Indian, and Pacific oceans as well as the east and west wind drifts of the Southern Ocean. The west wind drift and subtropical circulations meet and mix in the region of the subtropical convergence. The predominant movement of water throughout the area is zonal, upon which vertical and meridional movements are imposed. Flow is generally westward in the east wind drift near the Antarctic continent and eastward farther north, although meridional flow tends to predominate near the eastern and western edges of land masses. In addition, zonal flow in waters of the west wind drift near 45°S in the Pacific sector may be westward rather than eastward. The meridional and vertical circulation imposed on the zonal flow includes southward and northward components, respectively, in surface waters south and north of the Antarctic divergence, an upwelling region between the east and west wind drifts. The southward-flowing surface water contributes to a downwelling flow near the Antarctic continent, which moves into lower latitudes as bottom water. most of the northward-flowing surface water sinks in the region of the Antarctic Polar Front and slows into lower latitudes as intermediate water. A southward component of flow that occurs in deep layers between the intermediate and bottom layers is a source of the upwelling water at the Antarctic divergence. Surface waters of the west wind drift north of the Polar Front, which are warmer and more dilute than those farther south, also have a northward component of flow, whereas subsurface waters, above the intermediate layers, have an apparent southward component.

The major features of circulation contribute to the formation of five major zonal hydrologic regions connected by meridional and vertical components of circulation. The characteristics and limits of these regions are made somewhat obscure and variable by unique regional patterns of circulation and mixing, particularly near land masses. The regions are, from north to south, a warm water region associated with the subtropical gyres, the transitional region of the subtropical convergence, the subantarctic region (between the subtropical convergence and the Antarctic Polar Front), the region of the Antarctic Polar Front, and the Antarctic region.

A total of 84 lanternfish species have been shown to occur south of 30°S. These species comprise five major patterns of distribution associated with the major hydrologic regions within which occur variations in distribution strongly correlated with regional hydrology.

Thirty-nine species have been found to have distributions associated with the warm water region. These species constitute a significant proportion of a large, variable, and taxonomically very distinct warm water biogeographic complex.

Twenty-one species have been found to have distributions associated with the transitional region. This group, the transitional water biogeographic complex, includes four endemic genera. Approximately one third of the species are closely related to species in adjacent patterns of distribution, and approximately one fifth are known from similar latitudes in the northern hemisphere.

Ten species have been found to have distributions associated with subantarctic waters. This complex includes the southermost occurring species of three genera, the northernmost occurring species of two genera, and one species of a distinct subgeneric group which does not occur north of the transitional region.

Seven species have been found to have distributions associated with the region of the Antarctic Polar Front, and seven additional species have been found to have distributions associated with the Antarctic region. Juveniles of species in both groups are characteristically distributed farther north than are adults. These species, which constitute the Antarctic-Antarctic Polar Front complex, belong to three genera. Two of the genera are otherwise found only in subantarctic waters, and the third is found only in subantarctic and transitional waters, if we discount a markedly divergent warm water species.

The strong correlation between patterns of distribution and hydrology on major and regional scales, and additional considerations have led to the conclusion that the distributions of lanternfish populations are determined primarily by patterns of circulation and vertical distribution and by inherited tolerances for environmental factors. It is assumed that lanternfish evolution has been determined by the availablility of discrete hydrologic regions capable of maintaining passively drifting yet vertically mobile populations.

Consideration of the fossil record, paleoceanography, and generic endemism in the recent World Ocean has led to the following conclusions. The Myctophidae family probably originated during the late Mesozoic or early Cenozoic era, quite possibly from benthic ancestors which had an extended pelagic larval stage preadapted for maintenance in discrete patterns of circulation. The family had achieved considerable generic differentiation by the Eocene era and approached its present level of differentiation by Miocene. Conditions conducive to the evolution of recent faunas as defined by endemic genera were established in the Oligocene era and enhanced during the Miocene era. It is concluded that recent lanternfish complexes evolved from Antarctic-subantarctic, southern transitional, cold North Pacific, and warm water Tertiary faunas during the marked oceanographic fluctuations of the Pliocene and Pleistocene eras.

Appendix 1. List of Material Examined

Station number, number of specimens, and minimum and maximum standard lengths (in parentheses) are given. See section on materials and methods for additional explanation.

Protomyctophum anderssoni

Eltanin (USC). 99, 93 (19-47); 107, 1 (22); 109, 20 (17-54); 110, 4 (23-27); 123, 2 (27-44); 125, 5 (20-63); 132, 14 (39-65); 137, 2 (23-66); 141, 24 (14-44); 142, 3 (36-60); 143, 8 (19-54); 148, 55 (23-68); 149, 11 (17-60); 150, 8 (20-22); 154, 4 (20-50); 165, 1 (26); 235, 10 (22-62); 246, 8 (24-30); 247, 4 (24-59); 248, 2 (21-27); 252, 4 (22-57); 252, 54 (20-63); 259, 1 (23); 262, 1 (40); 282, 1 (44); 306, 7 (24-67); 310, 11 (32-58); 313, 7 (33-39); 318, 1 (43); 348, 5 (12-25); 354, 3 (15-16); 355, 10 (13-15); 359, 26 (14-63); 360 2 (37-53); 361, 2 (37-57); 364, 26 (25-62); 368, 16 (31-62); 375, 59 (11-61); 378, 17 (13-38); 379, 1 (38); 381, 7 (41-62); 382, 7 (39-65); 383, 7 (31-64); 388, 4 (34-60); 392, 6 (31-61); 386, 1 (63); 397, 4 (36-55); 449, 1 (48); 471, 1 (63); 563, 63 (11-58); 567, 1 (46); 571, 1 (20); 575, 1 (60); 580, 3 (53-59): 581, 4 (54-63); 588, 1 (43); 592, 1 (57); 593, 12 (13-52); 667, 2 (25-40); 670, 12 (24-56); 683, 7 (20-25); 687, 7 (23-38); 691, 1 (21); 692, 1 (57); 696, 2 (23-58); 697, 1 (67); 701, 1 (27); 702, 1 (38); 718, 15 (20-45); 719, 10 (22-60); 729, 3 (19-43); 737, 16 (19-66); 738, 85 (20-32); 767, 2 (32-35); 771, 3 (31-33); 775, 8 (28-53); 778, 2 (28-59); 781, 11 (34-69); 782, 8 (28-61); 785, 3 (24-62); 788, 4 (24-26); 789, 2 (41-58); 792, 7 (42-59); 793, 3 (41-60); 796, 2 (42-44); 802, 4 (42-61); 811, 7 (36-69); 832, 10 (19-66); 835, 26 (24-63); 836, 17 (23-60); 839, 35 (21-65); 846, 13 (30-60); 847, 11 (13-60); 849, 29 (30-62); 850, 7 (32-56); 854, 12 (45-65); 855, 5 (25-67); 858, 8 (31-60); 859, 28 (27-67); 864, 15 (27-65); 866, 20 (30-62); 867. 20 (31-65); 868, 21 (14-58); 874, 5 (38-60); 877, 1 (57); 878, 1 (50); 882, 22 (13-65); 883, 10 (14-65); 885, 12 (12-16); 886, 14 (15-62); 888, 23 (18-65); 889, 16 (33-57); 890, 3 (33-40); 891, 8 (41-63); 895, 16 (30-62); 898, 4 (30-64); 900, 32 (30-67); 901, 13 (28-66); 903, 7 (27-66); 904, 2 (60-63); 906, 13 (31-64); 911, 18 (53-64); 912, 14 (27-67); 914, 10 (32-61); 915, 4 (34-50); 917, 12 (29-36); 918, 2 (32-60); 922, 1 (31); 940, 1 (51); 943, 8 (41-57); 944, 3 (51-63); 946, 3 (55-63); 947, 2 (54-63); 949, 4 (54-66); 950, 3 (37-64); 952, 5 (54-69); 953, 7 (34-66); 957, 6 (14-19), 1100, 1 (21); 1107, 10 (21-56); 1113, 2 (17-18); 1114, 1 (35); 1119, 12 (41-61); 1120, 12 (40-65); 1121, 58 (19-65); 1132, 3 (18-56); 1133 2 (40-58); 1137, 3 (17-19); 1162, 12 (23-53); 1167, 6 (21-50); 1170, 15 (22-54); 1186, 9 (25-27); 1187, 3 (25-45); 1195, 2 (46-49); 1201, 87 (26-57); 1204, 68 (27-51); 1206, 38 (19-63); 1214, 73 (17-61); 1215, 120 (20-68); 1216, 1 (22); 1220, 22 (20-68); 1224, 15 (22-50); 1225, 1

(20); 1226, 3 (41-53); 1234, 8 (20-49); 1235, 1 (39); 1236, 28 (21-53); 1245, 2 (20-50); 1262, 11 (20-56); 1269, 27 (23-56); 1270, 21 (25-64); 1290, 19 (27-64); 65, (24-63); 1295, 1 (38); 1299, 28 (29-52); 1302, 1 (50); 1303, 44 (42-67); 1304, 19 (23-63); 1307, 132 (29-65); 1308, 19 (23-59); 1315, 5 (47-64); 1316, 7 (42-56); 1319, 11 (31-65); 1320, 9 (18-57); 1323, 24 (28-64); 1324, 34 (32-61); 1327, 13 (30-61); 1328, 1 (50); 1331, 2 (32-64); 1332, 6 (27-51); 1337, 14 (29-65); 1341, 12 (29-34); 1342, 17 (14-53); 1347, 21 (28-58); 1348, 8 (46-54); 1354, 26 (21-63); 1355, 16 (23-51); 1358, 22 (27-49); 1359, 20 (24-65); 1361, 10 (22-57); 1362, 14 (24-63); 1364, 14 (26-60); 1365, 15 (33-57); 1373, 110 (15-49); 1374, 69 (22-59); 1375, 14 (23-50); 1380, 9 (26-47); 1383, 9 (39-62); 1384, 6 (23-56); 1388, 10 (28-58); 1389, 10 (27-63); 1392, 23 (28-58); 1393, 9 (29-60); 1448, 5 (48-60); 1454, 9 (20-25); 1456, 18 (22-61); 1462, 53 (17-62); 1463, 9 (42-62); 1468, 7 (17-61); 1470, 14 (20-57); 1471, 2 (52-57); 1473, 9 (20-26); 1475, 8 (21-28); 1481, 4 (17-25); 1485, 1 (63); 1488, 4 (55-62); 1503, 7 (12-15); 1507, 13 (43-67); 1510, 10 (32-64); 1516, 2 (15-53); 1519, 7 (14-53); 1521, 8 (15-61); 1522, 51 (12-68); 1525, 1 (60); 1528, 6 (14-66); 1543, 1 (45); 1574, 3 (37-58); 1580, 4 (32-67); 1584, 6 (16-62); 1586, 8 (37-64); 1587, 4 (20-62); 1590, 23 (15-57); 1606, 4 (22-25); 1607, 58 (21-31); 1608, 19 (18-30); 1609, 16 (13-65); 1610, 4 (15-67); 1615, 25 (14-66); 1616, 16 (13-61); 1623, 40 (12-58); 1626, 24 (12-15); 1627, 6 (12-61); 1633, 7 (13-65); 1634, 67 (13-67); 1636, 16 (14-67); 1637, 13 (17-64); 1641, 12 (12-58); 1642, 3 (16-48); 1645, 49 (29-59); 1646, 3 (40-60); 1648, 52 (13-59); 1649, 169 (14-65); 1653, 36 (14-62); 1658, 11 (17-62); 1661, 8 (20-21); 1662, 3 (17-20); 1665, 29 (13-65); 1666, 19 (13-61); 1671, 34 (15-70); 1676, 91 (16-65); 1677, 27, (14-20); 1678, 183 (16-68); 1679, 59 (13-64); 1683, 38 (14-62); 1684, 17 (14-18); 1685, 2 (24-59); 1686, 304 (12-23); 1687, 31 (14-68); 1689, 106 (17-62); 1830, 3 (60-66); 1855, 6 (28-51); 1951, 1 (45); 1959, 2 (49-53); 1961, 1 (26); 1963, 1 (48); 1966, 4 (35-62); 1970, 15 (30-67); 1976, 62 (13-21); 1977, 81 (13-52); 1979, 15 (15-23); 1992, 7 (35-62); 2139, 2 (36-50); 2174, 24 (14-67); 2177, 21 (14-60); 2179, 12 (16-61); 2183, 15 (15-68); 2187, 64 (14-60); 2189, 14 (16-62); 2191, 31 (18-65); 2204, 7 (16-60); 2205, 13 (19-27); 2207, 8 (16-23); 2208, 8 (16-20); 2210, 138 (15-63); 2211, 38 (15-43); 2212, 47 (16-30); 2213, 56 (16-38); 2216, 8 (22-58); 2228, 1 (20); 2234, 12 (23-62); 2236, 10 (18-23); 2237, 144 (20-68); 2238a, 24 (14-63); 2238b, 3 (13-18); 2240, 29 (16-71); 2241, 54 (20-65); 2242, 47 (41-57); 2243, 24 (15-20); 2244, 115 (23-60); 2245, 15 (22-63); 2246, 7 (19-53); 2247, 3 (26-52); 2254, 38 (23-61); 2260, 20 (44-57); 2261, 8 (17-55); 2263, 65 (22-60); 2265, 22 (20-55); 2266, 12

(20-49); 2285, 1 (27); 2287, 1 (56); 2289, 3 (28-43); 2290, 55 (25-60); 2291, 34 (31-61); 2292, 31 (25-60); 2293, 14 (25-65); 2294, 41 (43-59); 2296, 3 (24-56); 2297, 3 (20-35); 2298, 8 (23-29); 2299, 33 (30-60); 2300, 26 (25-30); 2301, 1 (26).

Eltanin Cruise 33 (ANARE). 33-11, 1 (61); 33-12, 1 (55).

Eltanin (SOSC). 10, 1 (14); 14, 2 (16-18); 17, 1 (38); 16, 3 (40-43); 18, 2 (44-61); 20, 1 (45); 26, 2 (16-42); 47, 3 (41-64); 49, 7 (20-65); 52, 3 (37-57); 54, 3 (52-70); 56, 1 (59); 61, 3 (36-56); 62, 5 (13-40); 63, 6 (13-56); 69, 15 (12-60); 71, 1 (15); 74, 14 (15-20); 76, 1 (16); 79, 1 (14); 80, 1 (15); 82, 1 (26); 85, 1 (22); 92, 1 (21);, 111, 3 (28-31); 118, 7 (22-46); 124, 12 (17-26); 125, 3 (23-47); 144, 12 (21-61); 145, 32 (25-63); 147, 10 (25-63); 149, 8 (32-48); 151, 17 (22-57); 157, 2, (31-42); 162, 2 (46-52); 168, 2 (35-36); 328, 2 (32-35); 336, 1 (33), 340, 4 (34-35); 344, 2 (34-48); 348, 1 (28); 349, 2 (33-35);, 350, 34 (29-36); 351, 20 (30-61); 352, 18 (29-59); 356, 20 (25-63); 357, 13 (23-51); 362, 116 (33-39); 363, 2 (30-38).

Discovery. 64, 2 (18-21); 114, 1 (59); 239, 2 (25-27); 522, 1 (17); 658, 5 (37-42); 738, 8 (21-38); 742, 1 (28); 745, 2 (31-33); 951, 1 (27); 975, 1 (26); 1019, 1 (32); 1021, 1 (41); 1026, 1 (63); 1233, 1 (38); 1417, 1 (30); 1493, 1 (34); 1559, 23 (34-55); 1707, 1 (58); 1728, 1 (55); 1812, 2 (27-30); 1813, 1 (25); 1817, 2 (27-28); 1846, 1 (27); 1858, 4 (30-52); 2023, 1 (46); 2180, 1 (37); 2221, 1 (37); 2288, 1 (50); 2307, 1 (14); 2311, 1 (50); 2507, 2 (32-36).

British Museum of Natural History. 1901.11.8.2 (Victoria Land, Southern Cross), 1 (54); 1948.5.14.673 (west of Grahamland), 1 (61).

Protomyctophum tenisoni

Eltanin (USC). 108, 4 (35-38); 109, 1 (40); 131, 91 (34-44); 132, 1 (42); 133, 1 (43); 134, 1 (39); 141, 6 (38-43); 142, 1 (37); 143, 1 (41); 148, 1 (39); 150, 20 (35-40); 154, 3 (33-37); 175, 1 (14); 236, 1 (37); 246, 2 (40-41); 310, 1 (43); 313, 5 (17-18); 318, 2 (21-42); 319, 3 (43-46); 320, 13 (42-48); 336, 1 (33), 348, 5 (16-17); 354, 4 (16-17); 355, 7 (12-17); 359, 6 (17-49); 378, 22 (16-48); 379, 5 (17-20); 381, 2 (21-50); 388, 1 (18); 563, 3 (31-48); 567, 1 (32); 781, 2 (45-46); 782, 1 (42); 846, 1 (43); 852, 2 (17-18); 868, 2 (17-18), 874, 5 (16-17); 877, 9 (18-49); 878, 7 (17-21); 882, 2 (21-22); 883, 7 (18-20); 885, 24 (17-21); 888, 7 (17-51); 889, 3 (18-19); 957, 1 (20); 1100, 1 (38); 1114, 9 (36-41); 1119, 3 (36-38); 1120, 3 (37-39); 1121, 9 (36-40); 1142, 4 (39-45); 1170, 1 (38); 1203, 1 (38); 1206, 1 (39); 1214, 2 (37-44); 1241, 13 (40-44); 1247, 1 (40); 1294, 2 (39-41); 1295, 6 (40-46); 1298, 12 (41-46); 1299, 1 (40); 1302, 1 (41); 1303, 6 (37-44); 1304, 1 (38); 1306, 5 (37-41); 1307, 1 (37); 1315, 4 (40-50); 1316, 5 (39-44); 1324, 8 (37-44); 1328, 3 (41-42); 1333, 16 (40-48); 1336, 8 (40-45); 1337, 8 (42-48); 1348, 1 (43); 1362, 1 (45); 1379, 2 (46-49); 1396, 1 (17); 1439, 9 (33-38); 1443, 1 (32); 1448, 1 (34); 1454, 1 (34); 1462, 5 (33-36); 1467, 1 (36); 1470, 1 (33); 1471, 4 (36-39); 1485, 2 (41-43); 1488, 2 (39-41); 1501, 11 (20-49); 1503, 1 (24); 1521, 2 (19-22); 1587, 1 (33); 1606, 3 (27-32); 1608, 2 (31-32); 1622, 3

(18-38); 1623, 1 (37); 1642, 1 (31); 1646, 2 (25-35); 1648, 1 (32); 1649, 1 (31); 1653, 1 (42); 1679, 1 (32); 1682, 5 (34-38); 1684, 2 (33-41); 1689, 3 (30-36); 1830, 1 (26); 1841, 1 (17); 1971, 1 (19); 1976, 18 (19-23); 2179, 2 (35-37); 2187, 112 (26-41); 2189, 8 (31-40); 2205, 1 (29); 2209, 13 (35-40); 2210, 1 (33); 2211, 1 (42); 2212, 1 (40); 2213, 9 (35-41); 2234, 1 (39); 2244, 1 (40); 2254, 2 (35-40); 2298, 2 (41-42).

Eltanin (SOSC). 10, 4 (32-33); 11, 28 (21-35);. 14, 22 (27-33); 71, 4 (34-35); 74, 52 (30-36); 76, 7 (30-35); 79, 1 (24); 81, 1 (32); 145, 2 (44); 147, 1 (41); 147, 1 (41); 157, 1 (44), 168, 1 (14); 327, 5 (12-15); 328, 3 (16-17); 330, 7 (10-15); 332, 15 (9-16); 335, 10 (10-16); 336, 2 (14-17); 337, 2 (15-16); 340, 2 (15-17); 342, 4 (12-15); 345, 17 (12-18); 362, 3 (46-47).

Discovery. 65,1 (32); 107, 1 (47); 109, 3 (45-47); 207, 2 (34-35); 384, 1 (17); 493, 1 (46); 665, 1 (36); 666, 1 (34); 667, 1 (35); 820, 1 (17); 849, 1 (28); 1320, 1 (34); 1617, 1 (43); 1831, 2 (44-46); 2289, 1 (33); 2623, 1 (29).

William Scoresby. 203, 1 (35).

Protomyctophum sp. A

Eltanin (USC). 97, 2 (16-17); 99, 11 (16-22); 169, 6 (15-21); 175, 1 (14); 318, 1 (22); 459, 1 (69); 563, 4 (13-15); 593, 1 (73); 640, 1 (72); 667, 32 (15-75); 737, 1 (70); 738, 1 (67); 1050, 1 (72); 1072, 1 (70); 1121, 1 (71); 1170, 1 (18); 1186, 4 (16-19); 1269, 1 (21); 1285, 11 (19-22); 1379, 1 (26); 1380, 1 (27); 1480, 2 (18-20); 1524, 1 (67); 1590, 1 (72); 1642, 1 (73); 1777, 1 (17); 2183, 1 (73); 2211, 1 (81).

Eltanin (SOSC). 6, 1 (66); 14, 1 (60); 340, 1 (23); 345, 1 (24); 349, 5 (24-25); 350, 3 (21-24); 362, 2 (26-27).

Discovery. 36, 1 (64); 44, 2 (61-67); 718, 1 (37); 829, 3 (59-77); 1123, 1 (69); 1138, 1 (68); 1140, 1 (70); 1188, 1 (42); 1501, 1 (59); 1979, 1 (74); 1984, 1 (69).

Protomyctophum andriashevi

Eltanin (USC). 110, 1 (33); 123, 2 (35-36); 259, 1 (44); 235, 1 (43); 246, 1 (45); 318, 1 (39); 355, 1 (43); 364, 1 (45); 379, 1 (45); 392, 1 (43); 563, 1 (21); 667, 2 (35-39); 730, 2 (41-42); 785, 1 (45); 831, 1 (45); 867, 1 (43); 957, 2 (14-50); 1170, 1 (37); 1220, 2 (43-46); 1269, 1 (34); 1299, 1 (42); 1302, 2 (38-39); 1304, 1 (39); 1306, 3 (41-47); 1307, 2 (42-44); 1316, 1 (42); 1327, 1 (44); 1337, 1 (46); 1384, 1 (31); 1397, 2 (15-17); 1409, 16 (17-27); 1432, 1 (50); 1462, 3 (34-39); 1471, 3 (37-43); 1473, 3 (36-39); 1475, 13 (32-39); 1483, 1 (34); 1521, 2 (47-50); 1606, 3 (18-50); 1684, 1 (32); 1687, 3 (34-35); 1689, 1 (35); 1976, 4 (25-45); 2189, 1 (33); 2191, 2 (33-36); 2214, 1 (32); 2244, 2 (34-38); 2246, 2 (31-36); 2254, 3 (35-38); 2263, 1 (39); 2265, 5 (33-40).

Eltanin (SOSC). 327, 1 (39); 351, 1 (45); 354, 1 (27).

Discovery. 750 1 (45); 849, 1 (29); 990, 1 (44).

Protomyctophum sp. B

Eltanin (USC). 109, 1 (45); 149, 1 (44); 235, 1 (45); 310, 1 (18); 355, 1 (19); 375, 1 (61);

379, 4 (38-57); 381, 1 (50); 667, 1 (22); 775, 1
(56); 865, 2 (18-22); 866, 1 (18); 874, 1 (17);
888, 1 (34); 901, 1 (22); 1107, 2 (43-44); 1112,
5 (42-51); 1113, 2 (50-70); 1262, 1 (46); 1294, 2
(50-54); 1302, 4 (50-55); 1306, 1 (56); 1314, 1
(20); 1327, 2 (52-58); 1337, 2 (38-57); 1380, 1
(25); 1478, 2 (17-18); 1480, 1 (16); 1521, 3
(60-62); 1590, 1 (55); 1622, 1 (27); 1652, 2
(34-37); 1653, 1 (39); 2203, 1 (41); 2205, 2
(16); 2214, 1 (44); 2247, 1 (17).
Eltanin (SOSC). 111, 1, (15); 156, 1 (18); 327,
1 (23); 335, 2 (32-33); 348, 2 (22-53); 351, 2
(23-64).

Protomyctophum normani

Eltanin (USC). 1200, 1 (34); 1203, 1 (33);
1392, 1 (30); 1694, 4 (29-34); 1723, 1 (36);
1725, 1 (36); 1726, 1 (36); 1753, 1 (41); 1756, 2
(37-39); 1794, 1 (39); 1798, 3 (33-38); 1840, 1
(50); 2216, 1 (27); 2217, 3 (26-28); 2218, 1
(27); 2221, 2 (26-28); 2230, 1 (29); 2233, 1 (33).
Discovery. 946, 1 (37).
Scripps Institution of Oceanography SIO.61.45.
25c (Monsoon, 46° 53'42"S, 179°48'32"W),
4(33-33); SIO.61.38.25d (Monsoon 42°03.8'S,
70°30.9'E), 1 (52); South African Museum IK 7
(west of Slangkop), 2 (31-35).

Protomyctophum sp. C

Eltanin (USC). 97, 3 (17-40); 169, 15 (34-47);
667, 1 (38); 1287, 2 (41-48); 1288, 1 (46); 1405,
4 (42-46); 1409, 1 (46); 1695, 1 (51); 1728, 2
(54-59); 1754, 1 (65); 1777, 1 (62); 1832, 3
(33-41); 1834, 5 (34-40); 1835, 1 (34); 1838, 8
(33-64); 1841, 3 (31-35); 1976, 1 (47); 2234, 1
(54); 2236, 1 (54); 2244, 1 (55); 2246, 3
(55-59); 2253, 1 (56); 2288, 4 (51-54); 2289, 2
(58-59).
Eltanin (SOSC). 321, 2(46-64); 322, 1 (40).

Protomyctophum bolini

Eltanin (USC). 99, 14 (18-39); 109, 9 (16-55);
110, 2 (19-20); 123, 6 (19-50); 126, 1 (32); 131,
12 (37-49); 132, 11 (37-53); 133, 4 (37-43); 137,
2 (31-49); 138, 1 (46); 141, 4 (44-55); 142, 2
(46-52); 143, 61 (31-55); 148, 2 (34-51); 149, 3
(17-32); 153, 1 (53); 154, 5 (34-46); 169, 1
(21); 235, 11 (17-52); 247, 4 (16-54); 248, 3
(41-52); 252, 1 (22); 275, 1 (42); 285, 1 (54);
306, 4 (21-23); 310, 8 (19-51); 313, 2 (21-29);
318, 1 (54); 348, 4 (21-35); 359, 6 (20-53); 360,
4 (16-20); 361, 2 (20-23); 364, 7 (25-55); 368, 2
(20-37); 379, 29 (22-48); 381, 6 (25-52); 382, 5
(21-59); 383, 7 (20-52); 388, 5 (24-53); 392, 4
(22-23); 396, 1 (50); 404, 15 (38-48); 563, 24
(14-60); 575, 8 (35-45); 581, 2 (41-50); 588, 1
(44); 592,, 2 (35-44); 593, 8 (31-42); 597, 1
(39); 601, 2 (35-40);605, 2 (33-38); 611, 1 (41);
627, 1 (40); 632, 1 (37); 634, 4 (35-43); 640, 1
(35); 643, 3 (34-41); 654, 2 (36-42); 667, 8
(18-56); 670, 1 (27); 687, 1 (43); 692, 1 (56);
697, 7 (36-52); 702, 9 (35-53); 718, 2 (26-27);
729, 2 (38-40); 730, 31 (32-39); 737, 9 (23-60);
738, 5 (34-48); 767, 3 (21-22); 771, 1 (25); 775,
5 (18-53); 778, 3 (20-56); 781, 20 (18-55); 782,
3 (18-49); 788, 5 (37-40); 789, 4 (36-48); 792, 1
(40); 793, 6 (35-42); 796, 2 (40-41); 811, 1

(40); 812, 1 (38); 831, 62 (18-48); 832, 1 (21);
835, 7 (19-46); 836, 12 (19-55); 839, 8 (32-50);
846, 7)18-37); 847, 4 (20-53); 849, 6 (20-50);
850, 2 (22-24); 852, 3 (21-23); 854, 4 (37-40);
855, 2 (37); 857, 14 (37-56); 858, 6 (20-43);
864, 6 (35-42); 865, 15 (22-49); 866, 6 (21-45);
867, 6 (23-55); 868, 4 (20-43); 874, 2 (21-56);
878, 1 (22); 882, 3 (21-55); 883, 5 (25-52); 885,
34 (21-48); 886, 10 (21-52); 888, 8 (25-54); 889,
6 (24-47); 890, 19 (22-58); 891, 14 (24-53); 892,
29 (26-55); 895, 3 (30-35); 898, 3 (23-39); 900,
2 (23-40); 901, 2 (29-30); 903, 1 (42); 904, 1
(35); 906, 3 (27-37); 911, 60 (29-50); 912, 2
(34-40); 914, 1 (40); 915, 2 (39-46); 917, 12
(32-47); 918, 1 (41); 935, 3 (43-48); 943, 2
(40-45); 946, 3 (36-51); 949, 4 (42-47); 950, 6
(35-42); 952, 1 (40); 953, 8 (28-45); 957, 2
(34-52); 1029, 3 (44-48); 1038, 1 (42); 1044, 1
(35); 1050, 1 (46); 1072, 6 (36-52); 1107, 2
(32-50); 1112, 9 (29-52); 1113, 1 (32); 1114, 1
(52); 1120, 5 (34-50); 1121, 5 (29-37); 1133, 3
(33-49); 1137, 3 (39-48); 1142, 6 (43-50); 1162,
4 (36-40); 1163, 6 (34-52); 1170, 4 (17-28);
1185, 1 (19); 1201, 4 (17-38); 1204, 11 (16-36);
1213, 9 (36-55); 1214, 8 (38-42); 1215, 6
(29-54); 1220 61, (31-54); 1224, 1 (41); 1234, 2
(40-41); 1235, 1 (33); 1245, 23 (37-53); 1269, 13
(26-46); 1270, 3 (29-47); 1290, 2 (17-35); 1194,
8 (22-54); 1299, 2 (40-43); 1302, 18 (28-57);
1303, 3 (26-28); 1304, 9 (18-53); 1306, 15
(29-59); 1307, 2 (37-44); 1315, 24 (25-56); 1316,
28 (25-55) 1319, 1 (45); 1320, 2 (36-43); 1323, 6
(27-53); 1324, 6 (30-50); 1327, 4 (21-27); 1328,
1 (19); 1332, 3 (28-54); 1337, 21 (22-51); 1342,
16 (19-60); 1348, 45 (17-58); 1355, 33 (29-50);
1358, 1 (45); 1359, 9 (28-54); 1361, 3 (30-57);
1362, 1 (48); 1364, 7 (31-49); 1365, 5 (25-59);
1374, 9 (34-50); 1379, 1 (46); 1380, 1 (57);
1383, 2 (36-49); 1388, 1 (47); 1389, 1 (48);
1392, 2 (38-40); 1393, 2 (39-47); 1432, 1 (17);
1439, 3 (45-56); 1448, 1 (19); 1454, 4 (15-34);
1462, 2 (30-53); 1463, 3 (30-63); 1467, 2
(29-30); 1468, 2 (38-39); 1470, 4 (22-39); 1471,
1 (35); 1473, 6 (17-42); 1475, 2 (19-23); 1478, 4
(17-21); 1481, 2 (19-22); 1483, 1 (23); 1488, 2
(42-51); 1501, 1 (53); 1503, 1 (24); 1507, 1
(30); 1513, 1 (29); 1516, 4 (21-26); 1519, 4
(24-49); 1521, 16 (24-59); 1522, 2 (29-30); 1525,
1 (47); 1574, 6 (30-36); 1580, 1 (41); 1584, 1
(35); 1586, 1 (42); 1590, 1 (30); 1606, 20
(15-36); 1607, 6 (15-51); 1609, 5 (37-48); 1615,
7 (25-40); 1616, 2 (31-33); 1622, 24 (25-52);
1623, 3 (28-35); 1627, 1 (26); 1633, 6 (30-51);
1636, 1 (51); 1637, 3 (46-48); 1641, 2 (26-31);
1642, 8 (42-51); 1645, 4 (30-34); 1646, 2
(30-37); 1648, 13 (26-41); 1649, 8 (27-38); 1653,
1 (16); 1658, 2 (27-52); 1661, 2 (31-34); 1662, 1
(60); 1665, 1 (26); 1666, 3 (28-31); 1676, 7
(33-48); 1677, 24 (35-54); 1678, 3 (32-55); 1679,
5 (28-50); 1683, 8 (29-43); 1685, 1 (28); 1687,
20 (18-55); 1689, 5 (16-38); 1830, 1 (28); 1855,
1 (25); 1862, 1 (32); 1963, 2 (33-47); 1966, 5
(31-36); 1970, 4 (25-35); 1977, 2 (24-52); 1992,
3 (22-31); 2174, 7 (31-46); 2177, 5 (32-47);
2179, 5 (25-43); 2183, 3 (30-33); 2187, 8
(15-39); 2189, 9 (27-48); 2191, 1 (50); 2204, 3
(17-19); 2205, 35 (16-23); 2207, 2 (33-34); 2209,
1 (31); 2210, 6 (35-52); 2211, 25 (32-54); 2212,
10 (30-53); 2213, 57 (13-54); 2214, 28 (15-53);
2216, 4 (16-21); 2217, 24 (17-21); 2234, 2
(16-36); 2236, 24 (32-46); 2237, 3 (32-52);

2238b, 21 (33-51); 2241, 2 (36-49); 2242, 1 (35); 2243, 1 (36); 2244, 3 (32-37); 2246, 32 (18-43); 2247, 1 (16); 2254, 5 (22-54); 2260, 5 (37-42); 2262, 4 (27-43); 2263, 4 (27-43); 2266, 1 (23); 2268, 2 (17-21); 2289, 1 (39); 2290, 2 (20); 2294, 3 (37-42); 2295, 1 (34); 2296, 24 (25-55); 2298, 11 (41-53); 2299, 1 (23); 2300, 1 (41).

Eltanin (SOSC). 16, 3 (32-50); 17, 2 (43-49); 18, 2 (34-47); 20, 2 (43-44); 22, 2 (36-38); 26, 2 (46-47); 32, 1 (51); 41, 1 (42); 47, 10 (39-50); 49, 1 (44); 52, 1 (43); 54, 2 (38-46); 56, 3 (38-44); 62, 2 (38-46); 66, 1 (30); 69, 7 (35-43); 71, 4 (38-40); 76, 1 (51); 77, 2 (30-33); 79, 9 (25-34); 80, 13 (15-50); 82, 2 (18-19); 85, 2 (15-16); 111, 2 (18-19); 118, 12 (31-54); 125, 1 (42); 139, 6 (40-51); 140, 1 (35); 144, 7 (29-43); 145, 6 (20-51); 147, 4 (17-37); 149, 1 (23); 151, 12 (19-39); 156, 2 (19); 157, 3 (17-28); 159, 1 (26); 162, 1 (19); 327, 10 (17-29); 332, 2 (17-18); 335, 1 (17); 344, 2 (28-41); 345, 1 (20); 350, 1 (49); 351, 59 (17-47); 352, 7 (17-26); 354, 23 (18-51); 355, 98 (22-57); 356, 5 (28-43); 357, 5 (29-52); 358, 1 (39); 363, 6 (17-22).

Discovery. 72, 2 (38-42); 114, 1 (44), 671, 1 (58); 737, 1 (40); 751, 1 (38); 765, 1 (38); 849, 1 (19); 1019, 2 (23-49); 1033, 1 (48); 1063, 1 (43); 1064, 2 (49-50); 1233, 2 (47-49); 1434, 1 (40); 1448, 1 (56); 1455, 1 (41); 1493, 1 (28); 1559, 5 (28-50); 1775, 1 (19) 1779, 1 (42); 1798, 1 (46); 1809 1 (46); 1812, 2 (23-37); 1812, 1 (50); 1813, 1 (49); 1813, 5 (46-50); 1912, 1 (22); 2106, 1 (43); 2261, 1 (47); 2271, 1 (40); 2288, 1 (44); 2311, 1 (40); 2313, 1 (33); 2481, 2 (27-28); 2496, 1 (48); 2507, 1 (50); 2585, 1 (31); 2622, 2 (37-39); 2751, 1 (52).

Protomyctophum subparallelum

Eltanin (USC). 99, 2 (18); 169, 7 (19-22); 175, 5 (14-28); 213, 3 (24-27); 326, 1 (17); 849, 1 (23); 865, 3 (20-24); 883, 2 (25-28); 1269, 1 (23); 1286, 5 (19-26); 1287, 15 (15-31); 1288, 9 (24-28); 1337, 1 (26); 1355, 1 (25); 1694, 170 (15-34); 1696, 10 (16-31); 1700, 2 (18-25); 1719, 3 (17-24); 1723, 3 (18-20); 1724, 4 (16-20); 1726, 4 (16-21); 1727, 3 (20-22); 1728, 1 (20); 1731, 5 (19-21); 1734, 2 (19); 1736, 3 (19-20); 1741, 13 (18-32); 1753, 4 (34); 1755, 2 (17-20); 1774, 1 (32); 1777, 5 (17-22); 1781, 3 (20); 1792, 5 (22); 1797, 1 (21); 1798, 1 (24); 1802, 2 (19-22); 1803, 1 (24); 1811, 1 (19); 1824, 4 (27-30); 1829, 1 (28); 1834, 1 (28); 1835, 3 (27-28); 1841, 38 (25-29); 1842, 2 (27-29); 2205, 1 (31); 2216, 1 (15); 2217, 7 (18-20); 2221, 1 (19); 2226, 2 (21-32); 2227, 7 (20-26); 2229, 1 (22); 2232, 2 (19-20); 2269, 1 (21); 2271, 4 (22-24); 2273, 1 (21); 2279, 1 (21); 2283, 2 (19-21); 2302, 6 (23-26).

Eltanin (SOSC). 82, 1 (28); 92, 1 (28); 157, 1 (29); 168, 2 (22-31); 303, 1 (24); 306, 9 (16-29); 307, 1 (18); 314, 15 (18-30); 315, 8 (18-29); 316, 63 (18-33); 317, 15 (21-36); 321, 46 (16-25); 322, 12 (21-29); 327, 13 (18-34); 328, 2 (20-29); 332, 9 (16-30); 335, 132, 13 (19-35); 336, 7 (18-20); 337, 3 (18-27); 338, 3 (17-35); 341, 50 (18-31); 344, 1 (31); 345, 15 (17-36); 348, 12 (23-24); 351, 9 (21-26).

Discovery. 78, 1 (17); 85, 1 (21); 100, 1 (28); 100c, 2 (26-27); 101, 2 (27-28); 1806, 1 (25).

Protomyctophum parallelum

Eltanin (USC). 97, 5 (16-36); 99, 2 (22-40); 109, 81 (21-48); 110, 4 (21-37); 123, 6 (26-41); 125, 3 (29-43); 149, 2 (36-40); 165, 6 (28-38); 169, 296 (17-30); 175, 4 (15-21); 213, 9 (17-36); 215, 10 (18-38); 232, 2 (38-44); 235, 3 (26-44);; 248, 2 (24-27); 252, 2 (38-43); 253, 2 (40-43); 306, 30 (27-38); 310, 6 (18-40); 313, 5 (16-40); 348, 11 (17-41); 354, 1 (47); 355, 4 17-41); 359, 4 (30-49); 360, 3 (29-37); 361, 5 (19-39); 364, 1 (28); 375, 131 (17-48); 378, 42 (20-47); 381, 5 (25-46); 382, 14 (26-45); 388, 4 (31-35); 563, 5 (30-38); 670, 2 (40-46); 738, 1 (42); 741, 12 (16-36); 767, 2 (43-44); 771, 1 (31); 775, 19 (19-49); 778, 2 (23-33); 782, 2 (34-41); 788, 2 (39-40); 789, 1 (50); 832, 1 (41); 835, 6 (39-46); 836, 4 (36-42); 839, 4 (28-44); 846, 14 (21-47); 847, 4 (21-40); 849, 7 (30-40); 850, 3 (29-46); 852, 2 (21-30); 864, 8 (18-45); 866, 15 (24-43); 867, 6 (28-43); 868, 9 (26-40); 874, 2 (40-42); 877, 4 (18-43); 878, 2 (21-40); 882, 8 (19-33); 883, 8 (21-46); 885, 25 (20-42); 886, 4 (22-45); 888, 5 (21-40); 889, 3 (20-39); 890, 3 (23-42); 891, 2 (23-36); 892, 22 (20-44); 895, 1 (37); 898, 1 (35); 900, 1 (39); 901, 1 (39); 904, 1 (46); 906, 1 (20); 953, 2 (28-40); 957, 18 18-31); 1099, 1 (40); 1106, 6 (39-45); 1107, 17 (23-47); 1113, 17 (20-30); 1114, 2 (38-40); 1121, 1 (29); 1137, 1 (38); 1170, 5 (28-40); 1186, 3 (17-18); 1187, 1 (40); 1201, 3 (25-45); 1204, 13 (22-45); 1214, 1 (32); 1262, 1 (27); 1269, 7 (31-39); 1270, 4 (35-39); 1285, 4 (19-36); 1286, 5 (24-34); 1287, 18 (15-31); 1288, 3 (23-34); 1290, 5 (25-34); 1295, 1 (28); 1299, 3 (28-46); 1302, 27 (22-46); 1303, 4 (17-32); 1304, 18 (22-43); 1306, 27 (24-45); 1307, 8 (29-44); 1320, 1 (32); 1323, 3 (28-31); 1324, 1 (35); 1327, 4 (35-42); 1328, 3 (33-34); 1337, 2 (31-42); 1342, 29 (17-42); 1364, 1 (31); 1365, 2 (33-38); 1374, 4 (33-39); 1380, 12 (20-35); 1383, 1 (45); 1397, 1 (26); 1432, 12 (25-40); 1448, 2 (20-39); 1454. 8 (25-43); 1456, 2 (38); 1463, 2 (36-41); 1468, 1 (33); 1503, 1 (33); 1516, 2 (29-30); 1521, 6 (18-29); 1522, 5 (16-40); 1525, 3 (31-43); 1528, 1 (41); 1590, 47 (18-39); 1606, 27 (23-45); 1607, 10 (21-35); 1608, 4 (23-38); 1615, 1 (30); 1616, 1 (46); 1623, 34 (28-45); 1645, 3 (32-45); 1646, 4 (25-44); 1648, 5 (30-40); 1649, 1 (40); 1653, 2 (27-36); 1658, 3 (26-38); 1661, 13 (19-41); 1666, 1 (45); 1671, 1 (41); 1679, 1 (36); 1683, 1 (38); 1685, 1 (39); 1686, 4 (26-40); 1689, 1 (34); 1692, 4 (36-38); 1723, 2 (19-22); 1731, 1 (23); 1798, 5 (26-32); 1799, 3 (20-30); 1821, 1 (26); 1830, 2 (32-33); 1835, 3 (25-30); 1839, 7 (24-27); 1985, 1 (30); 1992, 1 (28); 2183, 1 (30); 2187, 6 (27-38); 2191, 2 (28-29); 2204, 5 (18-34); 2205, 1 (25); 2216, 1 (31); 2221, 3 (18-35); 2226, 1 (18); 2232; 1 (17); 2234, 1 (31); 2246, 1 (32); 2247, 4 (20-34); 2257, 3 (34-39); 2263, 1 (40); 2268, 2 (29-34); 2271, 2 (28-33); 2283, 1 (33); 2290; 1 (40); 2299, 2 (26-30); 2301, 1 (39).

Eltanin (SOSC) 62, 9 (17-34); 63, 1 (37); 71, 2 (18-19); 77, 3 (38-39); 80, 8 (28-42); 85, 2 (29-30); 92, 3 (15-35); 111, 4 (18-40); 149, 1 (16); 151, 4 (30-41); 156, 1 (27); 157, 7 (15-39); 159, 3 (16-17); 168, 7 (17-26); 301, 1 (33); 306, 5 (16-19); 314, 24 (15-18); 315, 25 (17-29); 317, 2 (17-39); 320, 1 (15), 321, 5 (17-40); 322, 27 (14-21); 323, 7 (15-26); 327, 21

(16-36); 328, 15 (16-37); 332, 2 (17), 335, 1 (17); 338, 9 (16-36); 340, 3 (31-32); 344, 2 (29-31); 348, 7 (17-19); 352, 5 (19-39); 354, 8 (28-37); 356, 4 (26-36); 363, 4 (16-33).

Discovery. 72, 2 (30-31); 1775, 1 (32); 1809, 1 (20); 1809, 1 (18); 2024, 1 (27); 2582, 1 (24).

Protomyctophum sp.D

Eltanin (USC). 1719, 1 (24); 1725, 3 (22-25); 1726, 2 (26-27); 1727, 1 (17); 1734, 2 (23-24); 1736, 1 (33); 1737, 1 (25); 1739, 4 (19-34); 1740, 1 (24); 1764, 1 (32); 1766, 1 (25); 1769, 1 (33); 1773, 1 (26); 1776, 1 (24); 1781, 1 (27); 1785, 3 (33-35); 1787, 3 (31-35); 1792, 12 (16-27); 1793, 2 (26); 1802, 3 (13-18); 1812, 2 (13-14).

Electrona antarctica

Eltanin (USC). 131, 1 (30); 132, 5 (35-65); 133, 3 (45-62); 134, 2 (45-62); 137, 7 (22-67); 138, 1 (59); 141, 34 (21-90); 142, 4 (39-65); 143, 13 (32-69); 148, 1 (36); 236, 16 (31-46); 247, 1 (59); 248, 2 (45-56); 259, 1 (23); 262, 1 (39); 274, 6 (40-77); 275, 16 (22-80)); 279, 5 (22-75); 280, 3 (21-72); 281, 43 (17-91); 282, 3 (69-79); 285, 38 (21-86); 292, 27 (19-92); 297, 5 (20-70); 302, 10 (34-77); 304, 24 (32-67); 306, 6 (20-52); 361, 2 (25-28); 364, 3 (37-49); 368, 40 (24-85); 375, 1 (24); 396, 57 (24-75); 397, 6 (24-70); 404, 3 (68-90); 414, 185 (29-99); 422, 62 (39-84); 431, 31 (47-88); 441, 4 (20-28); 448, 1 (73); 449, 5 (30-54); 478, 1 (57); 483, 2 (36-?); 495, 2 (71-92); 508, 1 (84); 526, 2 (62-80); 531, 1 (51); 535, 1 (51); 546, 2 (78-89); 550, 2 (67-80); 567, 5 (32-93); 570, 9 (51-71); 571, 4 (44-78); 572, 3 (18-45); 575, 4 (19-57); 580, 4 (18-99); 581, 9 (21-79); 588, 3 (?); 592, 3 (17-30); 593, 17 (20-63); 597, 1 (65); 601, 18 (38-85); 605, 2 (28-43); 611, 3 (54-84); 626, 3 (80-89); 627, 8 (22-88); 631, 1 (64); 632, 7 (22-95); 634, 8 (37-71); 635, 6 (23-76); 640, 3 (25-88); 643, 9 (17-83); 654, 1 (61); 667, 1 (19); 670, 1 (21); 683, 9 (18-89); 687, 1 (70); 691, 2 (22-55); 692, 2 (34-52); 696, 3 (45-85); 697, 10 (21-55); 701, 2 (51-52); 702, 29 (18-75); 703, 4 (66-69); 714, 5 (53-70); 718, 4 (19-30); 719, 11 (35-90); 726, 1 (52); 729, 6 (18-60); 730, 11 (17-99); 737, 10 (18-89); 738, 15 (21-91); 782, 1 (45); 785, 1 (62); 788, 20 (26-82); 789, 1 (32); 792, 11 (20-73); 793, 4 (34-63); 796, 6 (30-71); 802, 12 (24-66); 811, 19 (20-75); 812, 2 (21-65); 832, 2 (57-64); 835, 17 (26-66); 836, 3 (35-56); 839, 3 (35-74); 854, 10 (32-74); 855, 2 (57-74); 859, 2 (60-63); 864, 1 (37); 883, 2 (45-47); 885,, 3 (25-26); 888, 2 (49-55); 890, 22 (28-73); 891, 8 (35-64); 892, 19 (23-74); 895, 5 (27-73); 898, 8 (27-64); 901, 4 (40-68); 903, 9 (24-79); 904, 5 (23-72); 906, 8 (26-62); 909, 2 (56-62); 911, 10 (49-73); 912, 2 (46-67); 913, 1 (77); 914, 10 (25-40); 915, 4 (37-60); 917, 26 (23-80); 918, 3 (35-71); 919, 5 (27-78); 920, 4 (25-36); 929, 14 (22-42); 930, 3 (22-69); 933, 2 (21-62); 935, 8 (21-88); 936, 9 (23-71); 937, 1 (81); 940, 7 (24-69); 941, 3 (39-70); 943, 15 (23-90); 944, 4 (29-34); 946, 4 (27-73); 947, 10 (27-82); 949, 12 (24-78); 950, 4 (27-69); 951, 1 (60); 952, 7 (21-63); 953, 5 (37-48); 998, 11 (64-94); 1006, 12 (55-86); 1007, 2 (64-68); 1010, 3 (69-97); 1014, 1 (64); 1015, 1 (69); 1022, 5 (69-85); 1023, 12 (39-90); 1026, 37 (42-95); 1027, 3 (49-82); 1029, 14 (41-93); 1030, 9 (57-79); 1036, 16 (29-86); 1038, 9 (40-77); 1044, 22 (24-87); 1050, 3 (55-83); 1051, 5 (66-85); 1057, 17 (19-92); 1064, 18 (42-91); 1065, 15 (19-86); 1071, 8 (29-93); 1076, 10 (55-80); 1077, 12 (49-94); 1107, 1 (20); 1114, 14 (34-72); 1120, 1 (41); 1121, 3 (38-65); 1129, 3 (37-59); 1133, 2 (71-78); 1137, 6 (20-67); 1141, 14 (29-72); 1142, 45 (36-77); 1162, 10 (23-66); 1163, 19 (33-77); 1167, 1 (74); 1196, 1 (31); 1203, 1 (35); 1204, 1 (29); 1206, 3 (31-54); 1213, 3 (43-63); 1214, 17 (26-77); 1215, 3 (35-39); 1220, 12 (19-60); 1224, 12 (24-55); 1231, 13 (19-77); 1234, 16 (19-68); 1235, 4 (35-52); 1236, 20 (20-77); 1238, 51 (34-65); 1241, 37 (28-84); 1245, 72 (20-85); 1262, 1 (21); 1269, 1 (23), 1290, 1 (33); 1294, 2 (26-41); 1295, 9 (30-49); 1298, 1 (46); 1303, 1 (24); 1304, 16 (23-60); 1306, 11 (26-62); 1316, 15 (32-52); 1319, 3 (24-35); 1320, 3 (26-29); 1324, 2 (32-63); 1327, 1 (20); 1332, 3 (30-38); 1333, 17 (32-51); 1336, 13 (33-61); 1337, 2 (24-33); 1348, 34 (28-68); 1355, 43 (23-94); 1358, 19 (17-61); 1359, 17 (20-78); 1361, 4 (25-58); 1363, 4 (29-52); 1364, 3 (35-56); 1365, 1 (26); 1371, 7 (22-67); 1374, 34 (34-62); 1383, 1 (45); 1384, 4 (40-60); 1387, 1 (54); 1388, 3 (40-49); 1389, 3 (23-25); 1392, 11 (22-58); 1393, 4 (37-65); 1462, 7 (32-64); 1463, 8 (40-71); 1468, 3 (28-35); 1470, 2 (31-38); 1471, 5 (31-68); 1473, 3 (44-69); 1475, 1 (47); 1485, 9 (28-65); 1488, 21 (28-71); 1501, 2 (29-36); 1503, 1 (32); 1507, 1 (29); 1509, 1 (63); 1510, 17 (25-72); 1512, 9 (40-85); 1513, 1 (28); 1521, 1 (39); 1528, 3 (18-90); 1538, 3 (35-80); 1543 9 (39-92); 1546, 3 (55-69); 1552, 16 (44-87); 1559, 3 (29-32); 1564, 4 (33-67); 1568, 7 (35-82); 1574, 6 (21-61); 1576, 25 (22-95); 1580, 10 (23-88); 1584, 1 (25); 1609, 9 (30-57); 1610, 1 (51); 1615, 4 (49-74); 1616, 2 (42-52); 1622, 4 (41-65); 1623, 3 (31-53); 1633, 2 (42-56); 1634, 10 (22-70); 1636, 6 (26-74); 1637, 5 (25-65); 1641, 10 (37-68); 1642, 6 (29-67); 1648, 5 (30-44); 1649, 9 (21-67); 1657, 22 (27-57); 1658, 3 (31-39); 1661, 2 (19-23); 1665, 1 (28); 1666, 2 (38-41); 1671, 9 (29-76); 1672, 18 (27-40); 1676, 21 (28-73); 1677, 40 (30-69); 1678, 5 (16-50); 1679, 6 (27-63); 1683, 2 (45-53); 1684, 1 (36); 1686, 8 (29-43); 1689, 4 (18-74); 1855, 3 (38-46); 1862, 1 (65); 1865, 5 (21-70); 1868, 3 (30-69); 1936, 10 (20-89); 1942, 2 (37-73); 1947, 4 (42-67); 1955, 5 (29-55); 1959, 7 (30-75); 1963, 6 (24-72); 1966, 4 (36-67); 1970, 1 (42); 1976, 1 (22); 1977, 1 (39); 1992, 2 (33-60); 1993, 4 (30-61); 2111, 6 (27-68); 2114, 3 (33-47); 2122, 13 (22-90); 2130, 1 (29); 2133, 2 (28-39); 2136, 2 (22-46); 2139, 3 (38-45); 2168, 2 (38-71); 2174, 9 (32-72); 2177, 2 ((28-43); 2179, 7 (24-67); 2183, 12 (17-64); 2187, 9 (33-78); 2207, 9 (32-59); 2208, 8 (24-38); 2210, 46 (19-60); 2211, 31 (18-74); 2212, 16 (27-50); 2213, 33 (25-65); 2214, 1 (39); 2235, 17 (35-50); 2236, 24 (25-75); 2237, 9 (27-55); 2238a, 9 (33-55); 2238b, 35 (33-60); 2239, 34 (34-47); 2240, 12 (22-61); 2241, 14 (21-66); 2242, 1 (44); 2243, 1 (47); 2244, 2 (38-48); 2245, 1 (35); 2246, 3 (25-48); 2248, 2 (20-46); 2254, 2 (27-38); 2260, 20 (28-45); 2261, 17 (30-71); 2262, 27 (19-58); 2263, 10 (20-42); 2265, 3 (33-45); 2266, 1 (58); 2290, 2 (19); 2291, 1 (66); 2293, 7 (23-41);

2294, 7 (22-61); 2295, 1 (31); 2296, 17 (23-78); 2297, 22 (20-50); 2298, 37 (21-69); 2299, 4 (21-22); 2300, 3 (36-41).

Eltanin (SOSC). 11, 4 (26-60); 16, 1 (46); 18, 11 (27-69); 20, 43 (23-81); 26, 14 (26-68); 28, 9 (25-66); 29, 4 (29-34); 32, 1 (24); 33, 2 (51-53); 41, 14 (24-73); 43, 1 (40); 44, 10 (25-69); 47, 21 (22-80); 49, 16 (22-97); 50, 3 (61-74); 52, 3 (22-30); 54, 7 (18-86); 56, 8 (23-75); 58, 2 (30-56); 59, 1 (75); 60, 2 (21-54); 61, 6 (26-75); 62, 7 (25-82); 63, 7 (20-79); 68, 1 (27); 69, 3 (25-32); 71, 3 (57-67); 72, 1 (60); 74, 14 (27-62); 118, 8 (29-85); 124, 16 (42-73); 125, 23 (19-83); 129, 23 (19-83); 133, 6 (24-75); 139, 9 (47-87); 140, 25 (19-83); 144, 16 (19-71); 145, 9 (20-58); 147, 10 (24-69); 151, 9 (17-75); 354, 19 (23-62); 355, 7 (19-26); 356, 57 (19-92); 357, 20 (18-87); 358, 21 (31-63); 361, 7 (23-24).

Eltanin Cruise 33 (ANARE). 33-5, 7 (28-89); 33-5, 2 (48-75); 33-6, 5 (30-60); 33-7, 2 (37-41); 33-11, 9 (17-89); 33-12, 12 (22-87); 33-13, 2 (33-85); 33-15, 2 (24-27).

Discovery. 114, 19 (22-74) 116, 1 (68); 121, 1 (49); 197, 1 (75); 202, 1 (89); 357, 1 (30); 384, 1 (42); 407, 1 (78); 413, 1 (24); 461c, 1 (89); 462, 2 (58-66); 493, 2 (79-98); 500, 1 (92); 504, 1 (90); 520, 2 (36-52); 527, 9 (48-77); 643, 1 (72); 665, 2 (28-58); 737, 1 (46); 741, 2 (29-51); 743, 1 (68); 755, 1 (73); 757, 1 (40); 761, 1 (? 70); 763, 1 (?50); 765, 1 (73), 795, 1 (?55); 806, 1 (71); 850, 1 (51); 851, 3 (25-55); 852, 3 (32-35); 853, 2 (34-43); 856, 5 (27-42); 857, 1 (35); 858, 5 (35-58); 859, 1 (57); 861, 4 (27-38); 862, 2 (35-54); 863, 1 (45); 884, 2 (42-49); 884, 1 (?64); 885, 2 (43-45); 889, 1 (?35); 905, 1 (?40); 915, 1 (?50); 961, 1 (?80); 973, 1 (45); 975, 1 (54); 1017, 1 (43); 1015, 1 (45); 1052, 1 (85);1054, 1 (?85); 1061, 1 (?40); 1061, 1 (70); 1077, 1 (?95); 1125, 1 (47); 1140, 1 (?70); 1144, 1 (58); 1147, 3 (45-73); 1148, 1 (80); 1148, 1 (26); 1212, 1 (78); 1261, 1 (65); 1282, 2 (52-69); 1289, 2 (52-76); 1301, 1 (?70); 1314, 1 (29); 1327, 1 (?35); 1331, 1 (65); 1344, 1 (35); 1357, 7 (28-50); 1358, 1 (30); 1363, 3 (30-31); 1367, 1 (38); 1427, 1 (?24); 1434, 1 (51); 1437, 1 (?40); 1448, 1 (63); 1451, 1 (65); 1455, 1 (64); 1462, 1 (?30); 1501, 2 (50-83); 1513, 2 (40-50); 1515, 1 (?40); 1523, 3 (?30); 1543, 2 (?20-80); 1547, 1 (?30); 1547 2 (?-63); 1549, 1 (?70); 1550, 9 (31-60); 1551, 7 (27-37); 1559, 10 (21-84); 1624, 1 (33); 1628, 1 (38); 1636, 1 (63); 1694, 1 (30); 1701, 1 (76); 1705, 2 (29-52); 1707, 1 (34); 1711, 4 (59-71); 1723, 4 (28-55); 1728, 1 (62); 1778, 1 (23); 1778, 1 (36); 1779, 6 (21-48); 1780, 2 (30-63); 1784, 1 (61); 1786, 1 (57); 1788, 2 (34-38); 1792, 3 (25-90); 1793, 1 (98); 1812, 1 (55); 1812, 1 (41); 1813, 5 (24-36); 1815, 5 (32-53); 1815, 2 (22-45); 1820, 1 (?50); 1823, 2 (33-54); 1824, 1 (80); 1825, 1 (37); 1827, 3 (36-77); 1829, 1 (?35); 1845, 1 (?35); 1846, 1 (71); 1860, 1 (95); 1863, 1 (45); 1877, 3 (40-73); 1918, 1 (74); 1920, 1 (70); 1939, 1 (?80); 1943, 1 (?60); 1945, 1 (?80); 1947, 1 (84); 1965, 1 (79); 1979, 3 (61-81); 1983, 2 (44-46); 1989, 1 (44); 1990, 1 (95); 1993, 1 (89); 2000, 2 (34-57); 2004, 1 (73); 2007, 2 (42-69); 2009, 4 (27-63); 2012, 3 (39-60); 2014, 1 (?65); 2016, 1 (38); 2017, 2 (23-24); 2018, 1 (28); 2019, 2 (24-29); 2020, 1 (39); 2022, 1 (91); 2106, 5 (32-57); 2106, 2

(38-39); 2122, 2 (37-58); 2128, 1 (75); 2131, 4 (23-28); 2131, 4 (41-47); 2134, 1 (27); 2135, 2 (27-34); 2139, 6 (27-58); 2141, 2 (43-53); 2160, 2 (25-57); 2162, 1 (58); 2166, 1 (?30); 2171, 2 (22-51); 2221, 1 (29); 2224, 1 (?60); 2241, 1 (?50); 2244, 2 (30-35); 2250, 1 (41); 2261, 2 (?25-35); 2274, 1 (?35); 2280, 1 (72); 2293, 1 (22); 2297, 1 (87); 2307, 1 (38); 2310, 2 (37-38); 2313, 1 (84); 2316, 1 (40); 2318, 1 (52); 2318, 1 (70); 2334, 2 (?40-60); 2465, 1 (60); 2465, 4 (?20-35); 2474, 3 (35); 2479, 2 (?40-75); 2500, 1 (?100); 2535, 10 (27-65); 2567, 3 (37-54); 2571, 1 (?55); 2585, 1 (28); 2594, 1 (76); 2596, 3 (31-34); 2610, 1 (35); 2612, 1 (43); 2612, 1 (62); 2616, 4 (44-63); 2616, 2 (25-94); 2618 3 (38-67); 2618, 1 (40); 2620, 2 (28-29); 2620, 1 (64); 2645, 1 (?55); 2754, 1 (51); 2809, 1 (46); 2810, 1 (46), 2814, 1 (63).

William Scoresby. 30, 1 (60); 201, 1 (58); 343, 1 (75); 351, 4 (?30-50); 465, 1 (65); 535, 1 (90); 892, 1 (76); 894, 1 (65); 909, 1 (48); 910, 1 (63); 918, 1 (?78).

BANZARE [Norman, 1937]. 27, 1 (72); 32, 2 (55-56); 45, 1 (30); 96, 1 (58).

British Museum of Natural History. 1912-7-69 (Scotia 422: 68°32'S, 12°49'W), 1 (69); 1887-12-7-215-216 (Challenger 156, 157), 2 (57-72).

Electrona subaspera

Eltanin (USC). 98, 1 (45), 119, 1 (76); 165, 51 (25-84); 215, 2 (24-34); 313, 1 (62); 353, 1 (74); 375, 1 (81); 555, 2 (93-111); 563, 2 (23-24); 660, 7 (54-65); 667, 5 (20-69); 670, 1 (23); 741, 2 (30-32); 767, 1 (76); 885, 2 (85-86); 890, 3 (97-114); 957, 1 (54); 983, 6 (54-59); 987, 5 (54-93); 1106, 1 (72); 1167, 1 (29); 1170, 2 (25-26); 1179, 2 (68-89); 1270, 1 (74); 1285, 10 (22-26); 1297, 1 (59); 1328, 1 (24); 1337, 4 (26-78); 1339, 13 (62-87); 1342, 1 (20); 1348, 1 (26); 1409, 1 (102); 1454, 1 (56); 1463, 1 (70); 1606, 4 (52-66); 1617, 7 (86-97); 1645, 1 (88); 1648, 1 (33); 1655, 9 (37-96); 1662, 1 (45); 1678, 1 (81); 1689, 2 (64-66); 1693, 3 (61-73); 1698, 4 (30-62); 1700, 21 (23-25); 1701, 4 (22-24); 1724, 2 (24-25); 1725, 1 (26); 1726, 7 (22-27); 1727, 8 (21-25); 1728, 5 (23-25); 1730, 1 (102); 1731, 36 (19-26); 1734, 3 (23-25); 1739, 3 (22-25); 1740, 1 (21); 1753, 14 (23-27); 1755, 7 (23-25); 1756, 1 (27); 1761, 1 (JUV); 1777, 6 (23-27); 1781, 1 (24); 1798, 2 (23-27); 1799, 3 (23-25); 1802, 1 (25); 1823, 1 (27); 1841, 1 (38); 1966, 1 (100); 1970, 1 (117); 2187, 3 (24-71); 2202, 3 (52-78); 2214, 1 (73); 2216, 5 (23-25); 2217, 1 (29); 2228, 1 (34); 2232, 2 (24-26); 2247, 1 (27); 2270, 1 (26); 2281, 1 (29); 2283, 3 (25-28); 2285, 2 (23-24); 2286, 1 (69); 2300, 1 (95);

Eltanin (SOSC). 14, 3 (61-94); 80, 1 (67); 85, 2 (53-70); 92, 17 (21-26); 156, 1 (27); 162, 1 (24); 168, 1 (24); 255, 2 (81-83); 261a, 1 (78); 316, 2 (26-27); 317, 8 (24-34); 320, 1 (26); 321, 3 (26-30); 323, 1 (36); 326a, 1 (28); 327, 10 (26-32); 328, 54 (25-28); 330, 1 (58); 332, 1 (66); 336, 1 (27); 338, 25 (23-30); 340, 27 (24-28); 241, 9 (27-55); 344, 7 (23-29); 345, 3 (25-63); 348, 2 (27); 352, 1 (72).

Discovery. 71, 1 (26); 78, 1 (20); 104, 1 (32); 450, 1 (80); 716, 1 (32); 718, 2 (28-29); 1320, 1 (47); 1379, 1 (25); 1809, 2 (26-27); 1809, 1 (25); 1809, 1 (28); 2732, 4 (27-29);

2737, 3 (27-32); 2738, 2 (24-26); 2806, 1 (102); 2823, 1 (32); 2826, 1 (74).

British Museum of Natural History. 1845-8-5-45 (43°30'S, 123°E), 1 (93); 1874-3-27-3 (48°37'S, 84°16'W), 1 (85); 1913-12-4-206 (55°16'S, 120°03'W), 1 (96); 1969-1-24-572 (53°S, 100°W), 1 (52).

Electrona paucirastra

Eltanin (USC). 165, 85 (23-26); 167, 1 (60); 169, 5 (24-25); 173, 2 (55-57); 175, 3 (22-24); 213, 26 (23-60); 215, 10 (23-25); 326, 2 (33-46); 741, 2 (30-59); 1286, 3 (23-25); 1287, 3 (25-30); 1288, 4 (26-65); 1397, 1 (43); 1405, 3 (57-59); 1409, 1 (44); 1692, 2 (23-24); 1697, 2 (22); 1698, 4 (62-69); 1702, 4 (52-66); 1719, 2 (21); 1724, 2 (22); 1734, 2 (27-48); 1831, 2 (37-41); 1834, 1 (50); 1835, 3 (33-42); 1838, 1 (42); 1839, 1 (47); 1840, 1 (49); 1841, 4 (27-54); 1842, 8 (27-53); 2216, 6 (23-24); 2221, 12 (20-55); 2224, 12 (17-25); 2226, 1 (23); 2232, 1 (49); 2253, 5 (20-22); 2271, 2 (22-24); 2278, 1 (22); 2279, 1 (26); 2283, 21 (22-24); 2285, 2 (21); 2302, 2 (22-23).

Eltanin (SOSC). 303, 1 (23); 321, 1 (25); 322, 7 (25-26); 323, 4 (23-25).

Discovery. 78, 5 (21-22); 1680, 1 (27); 1803, 1 (25); 2147, 1 (31); 2155, 1 (43); 2803, 6 (38-59).

William Scoresby. 122, 1 (60).

BANZARE. 72, 1 (55).

British Museum of Natural History. 1876-3-4-78 (30°S, 170°W), 1 (55).

Metelectrona ventralis

Eltanin (USC). 326, 10 (17-46); 563, 7 (18-21); 667, 1 (17); 1286, 6 (18-39); 1287, 4 (25-28); 1700, 8 (18-20); 1701, 5 (18-67); 1723; 1724; 1734; 1753, 4 (17-20); 1758, 1 (22); 1823, 1 (50); 1831, 1 (60); 1838, 1 (48); 1840, 1 (55); 2224, 9 (19-22); 2226, 14 (19-22); 2227, 1 (21); 2232, 2 (19-20); 2273, 7 (19-22).

Eltanin (SOSC). 303, 8 (17-20).

Discovery. 267, 1 (22); 1370, 1 (20); 1617, 1 (29); 2036, 1 (19).

Scripps Institution of Oceanography. Anton Bruun cruise 18a, station 690b, 1 (67).

Metelectrona sp. A

Eltanin (USC). 175, 3 (18-20); 741, 5 (17-27); 1285, 5 (17-21); 1287, 2 (19-23); 1701, 2 (52-54); 1723; 1724; 1734; 1739, 2 (16); 1740, 1 (19); 1766, 2 (19-21); 1776, 1 (18); 1793, 1 (17); 1803, 1 (19); 1831, 10 (22-33); 1834, 3 (25-32); 1840, 7 (22-61); 2285, 2 (16-18).

Eltanin (SOSC). 92, 1 (20); 315, 8 (20-23); 317, 2 (22-27); 321, 4 (18-22); 322, 1 (32); 349, 1 (19).

Discovery. 78, 15 (17-18).

Metelectrona sp.

Eltanin (USC). 1693, 2 (20); 1719, 2 (16-21); 1727, 2 (16-17); 1728, 1 (18); 1741, 2 (20-22); 1755, 1 (19); 1761, 3 (16-17); 1781, 3 (16-19); 1985, 1 (17); 2216, 1 (26); 2253, 1 (20); 2268, 1 (19).

Eltanin (SOSC). 307, 1 (16).

Electrona carlsbergi

Eltanin (USC). 109, 2 (73-85); 122, 1 (61); 142, 1 (72); 143, 5 (75-85); 154, 1 (65); 235, 4 (66-79); 246, 11 (63-73); 306, 3 (68-82); 364, 9 (68-73); 478, 1 (70); 785, 1 (69); 836, 3 (62-66); 839, 1 (71); 864, 1 (69); 1142, 26 (61-72); 1162, 2 (60-61); 1163, 5 (63-68); 1213, 5 (62-66); 1214, 3 (50-66); 1215, 2 (63-64); 1220, 2 (58-62); 1236, 2 (61-82); 1245, 8 (63-68); 1247, 13 (60-71); 1302, 1 (80); 1303, 11 (58-62); 1304, 2 (64-65); 1319, 1 (60); 1324, 2 (64); 1355, 1 (42); 1359, 2 (67-72); 1397, 1 (11); 1405, 1 (23); 1407, 7 (17-23); 1409, 65 (13-26); 1427, 36 (13-21); 1428, 1 (20); 1432, 1 (17); 1462, 1 (66); 1467, 3 (65-73); 1468, 1 (66); 1470, 3 (62-65); 1484, 2 (63-65); 1485, 3 (67-70); 1491, 6 (60-71); 1507, 3 (70-77); 1509, 1 (76); 1538, 41 (66-74); 1580, 3 (71-76); 1609, 3 (14-20); 1633, 4 (69-80); 1634, 1 (63); 1636, 1 (69); 1676, 1 (72); 1677, 45 (60-79); 1678, 1 (71); 1695, 2 (35-39); 1839, 2 (14-16); 1841, 2 (12-18); 1842, 1 (13); 1970, 1 (60); 1972, 2 (12); 1979, 2 (12); 1982, 3 (13-17); 1984, 1 (23); 2174, 1 (72); 2179, 1 (60); 2191, 1 (21); 2208, 1 (63); 2211, 1 (61); 2216, 1 (37); 2218, 1 (37); 2227, 1 (33); 2238b, 2 (68-69); 2244, 1 (92); 2245, 1 (82); 2248, 1 (38); 2260, 1 (67); 2268, 3 (34-37); 2286, 3 (35-53); 2292, 1 (91); 2296, 1 (69); 2298, 2 (80-89); 2299, 1 (86).

Eltanin (SOSC). 62, 1 (73); 66, 4 (62-72); 69, 1 (70); 147, 1 (67); 315, 2 (11); 323, 1 (48); 348, 1 (47); 349, 6 (40-48); 350, 1 (52); 351, 1 (49); 362, 4 (70-87).

Discovery. 893, 1 (29); 1161, 1 (90); 1689, 2 (18); 2156, 1 (12); 2313, 1 (78).

BANZARE. 69, 1 (31).

British Museum of Natural History. 1948-5-14-21 (60°22'S, 102°05'E), 1 (101).

Electrona rissoi

Eltanin (USC). 2223, 1 (62).

Discovery. 87, 1 (14); 101, 2 (61-71); 285, 3 (21-23); 1736, 1 (13).

Benthosema suborbitale

Eltanin (USC). 1823, 1 (23).

Discovery. 100b, 1 (27); 257, 1 (29); 419, 1 (22); 435, 2 (19-27); 437, 5 (15-30); 439, 1 (17); 527, 1 (13); 701, 1 (19); 704, 1 (31); 875, 1 (26); 1371, 3 (15-29); 1372, 2 (24-25); 1568, 1 (27); 1582, 3 (24-27); 2067, 6 (19-25).

William Scoresby. 1031, 1 (22).

Diogenichthys atlanticus

Eltanin (USC). 1738, 3 (22-23); 1768, 6 (15-16); 1786, 1 (15); 1788, 2 (15-23); 1808, 8 (16-24); 1809, 15 (14-15); 1810, 6 (14-15); 2271, 8 (17-20); 2278, 65 (16-21); 2279, 9 (16-29); 2280, 5 (18-19); 2281, 1 (19).

Eltanin (SOSC). 306, 1 (22).

Discovery. 100c, 4 (16-24); 285, 2 (13-20); 287, 1 (17); 288, 1 (15); 289, 1 (19); 296, 2 (20-21); 413, 1 (25); 429, 1 (17); 433, 1 (19); 439, 2 (14-19); 673, 1 (19); 692, 1 (16); 695, 1 (13); 701, 4 (12-18); 713, 1 (25); 875, 1 (15); 878, 1 (16); 1760, 1 (18).

William Scoresby. 600, 1 (14); 604, 1 (15).

Hygophum hygomi

Eltanin (USC). 2273, 5 (16-21); 2278, 1 (21).
Discovery. 415, 1 (16); 419, 3 (15-22); 433, 1
(16); 437, 3 (15-17); 438, 3 (17-19); 440, 3
(15-18); 441, 1 (40); 1612, 1 (19); 2484, 1 (18).

Hygophum hanseni

Eltanin (USC). 1401, 1 (22); 1402, 6 (14-20);
1405, 2 (17-27); 1525, 18 (17-27); 1704, 1 (46);
1723, 1 (19); 1726, 2 (18-23); 1727, 1 (37);
1728, 2 (20-40); 1731, 4 (28-33); 1734, 1 (30);
1738, 1 (38); 1753, 6 (31-46); 1755, 1 (39);
1756, 1 (18); 1773, 2 (16-17); 1774, 1 (17);
1777, 4 (15-37); 1793, 1 (40); 1794, 1 (18);
1798, 6 (20-44); 1799, 4 (32-42); 1803, 7
(17-39); 1817, 14 (16-30); 1820, 5 (15-22); 1821,
3 (16-21); 1823, 6 (15-30); 1824, 5 (15-20);
1825, 2 (18-22); 1831, 69 (14-30); 1832, 6
(16-26); 1834, 2 (24-25); 1984, 1 (23); 1985, 9
(21-39); 2222, 1 (25); 2271, 3 (21-28); 2278, 2
(17-21).
Eltanin (SOSC). 315, 6 (15-46); 317, 3
(37-41); 320, 1 (17); 321, 6 (16-41); 322, 1
(35); 323, 2 (18-21).
Discovery. 71, 2 (26-29); 439, 1 (17); 2808, 1
(32).
Anton Bruun cruise 13 [Craddock and Mead,
1970]. 17, 2 (23-27); 41, 4 (14-26).

Hygophum bruuni

Eltanin (USC). 175, 3 (13-15); 177, 1 (20);
213, 2 (13-14); 215, 1 (22); 326, 40 (15-32);
741, 4 (15-19); 1286, 11 (15-44); 1287, 12
(15-18); 1288, 1 (42).
Eltanin (SOSC). 88, 1 (41); 302, 2 (12-14);
303, 4 (11-42); 307, 3 (13-24); 315, 6 (15-46);
317, 3 (37-41); 320, 1 (17); 321, 6 (16-41); 322,
1 (35); 323, 2 (18-24).
William Scoresby. 599, 1 (15); 600, 1 (16);
601, 6 (15-20); 604, 2 (13-15); 605, 5 (13-14).
Anton Bruun cruise 13 [Craddock and Mead,
1970]. 2, 92 (14-50); 6, 29 (14-36); 7, 1 (25);
8, 1 (35); 10, 15 (17-29); 16, 10 (18-29); 40, 1
(16); 41, 1 (17); 43, 3 (14-22); 44, 4 (14-24);
45, 6 (14-21); 46, 7 (15-50); 47, 3 (20-49); 49,
12 (22-42); 50, 60 (14-51); 52, 24 (15-50); 53,
18 (14-50); 54, 10 (13-41); 57, 22 (15-44); 58, 3
(26-34); 59, 6 (13-42); 62, 10 (14-41). Cruise
18a: LWK 66-45, 6 (35-42).

Hygophum macrochir group

Eltanin (USC). 1402, 1 (30); 1808, 1 (28);
1809, 2 (18-20); 1811, 1 (51); 2271, 3 (32-34);
2278, 5 (20-40); 2279, 3 (21-38); 2280, 1 (22).
(SOSC) cr. 25: 307, 3 (13-24).
Anton Bruun cruise 13 [Craddock and Mead,
1970]. 17, 3 (23-25); 18, 9 (17-27); 19, 3
(32-47); 20, 11 (21-39); 21, 7 (19-49); 23, 4
(18-52); 26, 2 (28-53); 27, 1 (23); 28, 1 (24).

Myctophum phengodes

Eltanin (USC). 1287, 1 (19); 1719, 1 (36);
1724, 2 (31-37); 1738, 2 (30-33); 1764, 2
(24-26); 1765, 5 (22-35); 1769, 1 (36); 1786, 1
(39); 1788, 1 (26); 1807, 2 (20-77); 1809, 1

(18); 1810, 2 (17-18); 1812, 4 (18-19); 2278, 1
(33); 2279, 1 (30).
Eltanin (SOSC). 306, 1 (25).
Discovery. 87, 1 (23); 413, 1 (28); 1372, 2
(17-21); 1744, 1 (23); 1760, 1 (58); 2032, 1
(83); 2868, 1 (20).
British Museum of Natural History. 1926-6-30-9
(Lord Howe Island), 1 (39); 1935-5-6-1 (Off Cape
Point), 1 (85).

Symbolophorus sp. A

Eltanin (USC). 165, 1 (43); 1655, 2 (103-105);
1726, 1 (76).
Eltanin (SOSC). 255, 1 (124); 317, 1 (49);
332, 1 (63); 340, 1 (136); 345, 1 (115); 347, 1
(124); 348, 1 (120).
British Museum of Natural History. 68-22-33
(44°42.5'S, 91°20'W), 1 (99).
Scripps Institution of Oceanography. SIO
58-255 (43°50'S, 105°30'W), 1 (86); SIO
58-260 (38°46'S, 83°20'W), 1 (67); Piquero
III station 41 (48°42.5'S, 91°20'W), 3
(92-154).

Symbolophorus sp. B

Eltanin (USC). 97, 4 (20-35); 99, 4 (22-24);
110, 1 (22); 165, 33 (20-41); 167, 5 (41-83);
173, 4 (51-108); 175, 4 (21-23); 177, 1 (?90);
178, 3 (87-122); 181, 1 (91); 213, 20 (21-49);
215, 2 (21-23); 326, 3 (28-38); 563, 3 (21-33);
741, 15 (20-30); 1286, 2 (75-79); 1698, 2
(33-37); 1701, 1 (78); 1702, 2 (33-40); 1827, 2
(30-46); 1831, 23 (23-60); 1834, 1 (43); 2227, 1
(59); 2232, 2 (37-43); 2233, 1 (44); 2281, 1 (68).
Discovery. 7, 1 (67); 718, 1 (58).
William Scoresby. 1070, 1 (87).
Anton Bruun cruise 13 [Craddock and Mead,
1970]. 4, 1 (28); 16, 1 (24); 24, 2 (34-36); 47,
9 (35-54); 48, 1 (48); 49, 28 (25-26); 50, 27
(23-75); 52, 4 (24-44); 53, 9 (27-67); 54, 2
(41-42).
Scripps Institution of Oceanography. SIO 61-79
(49°26.5'S, 132°18.4'E), 14 (110-135); Anton
Bruun cruise 5, station 308a, label 2567
(41°01'S, 75°E), 1 (95); Anton Bruun cruise
18a, station 66-32 (35°24'S, 73°18'W), 1
(83); SAM (West of Cape point), 1 (90).

Symbolophorus sp. C

Eltanin (USC). 1433, 1 (85).
Discovery. 1281, 1 (67).
Dana. 3644 (44°40'S, 173°39'E), 2 (78-79).
Scripps Institution of Oceanography. Dominion
Museum (New Zealand) 2144 (45°38'S, 171°12'E),
6 (73-89).

Symbolophorus sp. D

Eltanin (USC). 742, 1 (22); 1287, 2 (23-26);
1761, 1 (23); 1766, 9 (24-27); 1772, 1 (67);
1773, 1 (80); 1784, 1 (101); 1786, 14 (24-27);
1799, 2 (22-25); 1802, 1 (28); 1807, 2 (27-28);
1810, 3 (25-27); 1811, 3 (30-39).
Eltanin (SOSC). 304, 1 (99).
Anton Bruun cruise 13 [Craddock and Mead,
1970]. 2, 10 (35-70); 7, 1 (26); 10, 2 (26-27);
16, 9 (24-59); 17, 17 (22-56); 18, 2 (24-47); 20,
2 (22-26); 24, 1 (25); 29, 1 (28); 30, 1 (31);

40, 2 (24-25); 43, 31 (24-48); 46, 29 (28-79); 47, 26 (28-90); 49, 3 (22-26); 50, 4 (27-65); 53, 1 (26).

British Museum of Natural History. 68-6-22-35 (South Pacific), 2 (42-71).

Scripps Institution of Oceanography. SIO 57-161 (24°54'S, 71°26'W), 49 (32-65); SIO 65-641-25a (Isla Robinson Crusoe), 1 (53); SIO 65-663 (33°31'S, 75°18'W), 10 (83-113); Anton Bruun cruise 18a, station 66-45 (32°19'S, 71°46'W), 1 (76).

Symbolophorus spp.

Eltanin (USC). 1822, 1 (87); 2223, 1 (62).

Discovery. 71, 3 (39-44); 716, 1 (50); 1566, 2 (50-67); 1604, 1 (58); 1768, 1 (108).

William Scoresby. 1100, 4 (36-68).

British Museum of Natural History. 1922-1-13-61 (Uzumbi, Natal), 1 (96); 1926-6-30-6-8 (Lord Howe island), 3 (84-101); 1948-1-12-1 (type of Myctophum boops Richardson), 1 (95).

Scripps Institution of Oceanography. SIO 61-123 (36°38'S, 95°29.6'E), 1 (70); Sam (West of Cape Point), 2 (81-102).

Zoological Museum, University of Copenhagen. P2329227 (type of Myctophum humboldti barnardi Tåning), 1 (95).

Loweina interrupta

Eltanin (USC). 1704, 2 (31-32); 1766, 1 (23); 1811, 1 (27); 2222, 9 (29-34); 2223, 3 (31-32); 2271, 2 (27-32); 2281, 3 (28-30).

Discovery. 87, 1 (33); 257, 1 (27); 1748, 1 (29).

Gonichthys barnesi

Eltanin (USC). 1401, 1 (34); 1719, 1 (27); 1724, 6 (26-42); 1758, 2 (28-30); 1774, 1 (42); 1781, 1 (39); 1786, 1 (22); 1802, 3 (23-31); 1803, 2 (30-40); 2273, 13 (23-41); 2278, 1 (30)

Eltanin (SOSC). 307, 2 (17-20); 308, 1 (38).

Discovery. 87, 1 (19); 247, 1 (33); 673, 1 (26); 713, 1 (21); 878, 1 (32); 1372, 1 (38); 1609, 1 (27); 1737, 1 (33); 1769, 1 (30); 2030, 1 (31).

British Museum of Natural History. 1868-6-22-35 (South Pacific), 12 (23-51); 1879-5-14-602-606 (Challenger 597), 5 (29-44); 1889-7-20-6-7 (38°07'S, 94°04'W), 2 (22); 1904-5-28-1-9 (45 mi east by north from Cape Point), 9 (45-50); 1912-11-28-81-85 (Challenger 597), 13 (24-40).

Notolychnus valdiviae

Eltanin (USC). 1788, 1 (17); 1808, 1 (13); 1823, 1 (23).

Discovery. 284; 701; 702; 711; 1586; 2067; 2073; 712.

Lampadena speculigera

Eltanin (USC). 1402, 1 (20); 1710, 1 (66); 1820, 1 (109); 2222, 2 (18-21).

Lampadena notialis

Eltanin (USC). 1830, 1 (66); 1841, 1 (105).

Lampadena dea

Eltanin (USC). 1739, 1 (59).

Taaningichthys bathyphilus

Eltanin (USC). 947, 1 (67); 1724, 1 (60).

Bolinichthys supralateralis

Eltanin (USC). 1807, 1 (23); 2226, 1 (54); 2231, 1 (55).

Lepidophanes guentheri

Discovery. 66, 1 (28); 69, 1 (42); 76, 1 (35); 240, 2 (36-40); 241, 2 (27-44); 242, 2 (43); 281, 1 (48); 284, 3 (40-66); 286, 4 (41-49); 288, 5 (32-54); 289, 2 (31-32); 296, 1 (23); 672, 1 (50); 673, 2 (31-40); 688, 2 (32-37); 689, 1 (54); 692, 3 (20-52); 694, 1 (40); 698, 1 (51); 701, 4 (41-70); 702, 1 (24); 704, 3 (39-56); 710, 25 (17-52); 711, 6 (32-55); 712, 4 (42-50); 713, 3 (21-50); 714, 10 (44-51); 717, 4 (46-51); 718, 3 (36-37); 1598, 2 (63-65); 2057, 1 (45).

Ceratoscopelus warmingi

Eltanin (USC). 1710, 1 (27); 1723, 1 (44); 1724, 1 (30); 1736, 2 (23-33); 1737, 1 (35); 1738, 2 (26-39); 1764, 1 (33); 1766, 58 (20-68); 1765, 4 (21-42); 1768, 1 (23); 1772, 6 (23-49); 1773, 1 (68); 1781, 1 (38); 1786, 45 (22-68); 1787, 19 (21-26); 1788, 112 (21-59); 1792, 1 (23); 1793, 1 (69); 1794, 2 (65-67); 1804, 1 (67); 1809, 1 (22); 1810, 6 (21-24); 1811, 9 (22-51); 1812, 4 (19-38); 1823, 1 (35); 1831, 1 (38); 2222, 3 (19-21); 2223, 2 (21-53); 2224, 5 (16-24); 2271, 2 (23-24); 2278, 13 (22-45); 2280, 2 (29-49); 2281, 2 (33-34).

Eltanin (SOSC). 306, 1 (28); 307, 1 (26).

Discovery. 104, 1 (50); 270, 1 (50); 285, 1 (64); 286, 1 (44); 287, 1 (31); 288, 1 (53); 294, 1 (18); 296, 1 (18); 297, 1 (17); 407, 1 (31); 413, 1 (25); 679, 1 (22); 683, 2 (31-63); 692, 1 (23); 698, 1 (64); 699, 1 (19); 701, 2 (28-30); 703, 2 (28-53); 711, 1 (24); 1371, 1 (21); 1568, 3 (23-24); 1758, 1 (39); 1759, 1 (32); 1768, 1 (20); 1770, 2 (20-23); 1803, 1 (75); 2057, 1 (40); 2059, 4 (19-20); 2061, 1 (22); 2064, 3 (25-26); 2067, 3 (17-22); 2152, 1 (59); 2867, 1 (32).

British Museum of Natural History. 1876-3-4-83 (South Pacific), 1 (52); 1935-7-6-2 (Off Cape Point), 1 (30); 1939-5-24-509-512 (Murray 95), 2 (17-19); 1939-5-24-541 (Murray 131 c-d), 1 (15); 1969-6-26-657-665 (Rosaura 14, 33, 45), 9 (18-23).

Lampanyctus niger-ater complex

Eltanin (USC). 1397, 1 (118); 1402, 2 (55-93); 1704, 1 (76); 1710, 5 (69-96); 1835, 1 (42); 2222, 1 (62); 2224, 1 (63); 2226, 2 (25-96); 2232, 1 (121); 2280, 1 (34).

Discovery. 81, 2 (57-58); 241, 1 (73); 257, 1 (70); 1565, 1 (67); 1565, 1 (67); 1604, 1 (40); 1763, 4 (66-85); 1805, 2 (31); 1807, 1 (74).

Lampanyctus achirus

Eltanin (USC). 97, 3 (121-154); 99, 3 (121-132); 109, 6 (101-128); 110, 1 (146); 123, 1

(102); 125, 1 (130); 132, 8 (97-138); 133, 1 (112); 142, 2 (100-102); 143, 7 (95-132); 148, 2 (90-103); 149, 1 (119); 165, 15 (29-151); 175, 1 (154); 215, 7 (33-118); 232, 1 (116); 235, 2 (113-126); 247, 3 (106-133); 248, 2 (111-126); 252, 1 (93); 253, 11 (66-125); 259, 1 (109); 275, 1 (116); 279, 1 (125); 292, 1 (126); 297, 1 (104); 306, 5 (113-120); 310, 1 (128); 313, 1 (153); 326, 1 (124); 348, 4 (93-142); 359, 3 (108-119); 360, 1 (83); 368, 3 (112-125); 374, 2 (122-124); 382, 3 (111-125); 383, 4 (60-135); 388, 3 (117-143); 392, 1 (124); 396, 1 (111); 667, 2 (57-126); 670, 2 (122-123); 683, 1 (112); 718, 2 (127-131); 737, 4 (120-166); 738, 3 (103-128); 767, 3 (91-120); 771, 2 (109-140); 779, 1 (116); 782, 3 (94-148); 785, 1 (121); 802, 3 (122-136); 812, 2 (126-130); 832, 1 (129); 835, 4 (84-129); 836, 3 (91-128); 839, 5 (74-108); 846, 4 (82-138); 847, 3 (79-134); 849, 7 (77-127); 852, 1 (122); 854, 6 (97-143); 858, 2 (122-125); 859, 3 (112-118); 864, 2 (86-123); 866, 14 (70-138); 867, 1 (128); 874, 3 (120-134); 877, 2 (83-124); 882, 2 (60-120); 883, 4 (104-123); 886, 3 (133-135); 889, 1 (103); 890, 2 (55-110); 891, 2 (98-116); 892, 2 (103-108); 895, 1 (114); 900, 5 (107-112); 906, 4 (99-112); 912, 1 (97); 914, 2 (120-125); 935, 1 (130); 940, 1 (109); 943, 1 (101); 946, 1 (105); 947, 1 (112); 949, 3 (109-135); 952, 1 (133); 953, 3 (114-129); 1099, 10 (74-125); 1106, 7 (75-141); 1107, 22 (40-139); 1112, 5 (108-120); 1113, 1 (121); 1114, 1 (77); 1120, 3 (115-127); 1121, 2 (90-115); 1132, 1 (98); 1133, 3 (98-120); 1137, 2 (117-122); 1162, 6 (91-139); 1167, 1 (100); 1170, 7 (75-137); 1185, 3 (77-139); 1187, 4 (99-122); 1201, 4 (56-98); 1204, 6 (96-134); 1206, 5 (75-128); 1214, 4 (93-106); 1215, 4 (77-108); 1220, 1 (93); 1262, 4 (69-114); 1269, 2 (65-120); 1270, 4 (50-105); 1285, 2 (30-33); 1286, 1 (35); 1290, 3 (111-124); 1294, 8 (81-142); 1299, 4 (92-123); 1302, 2 (80-99); 1303, 8 (99-125); 1304, 3 (43-132); 1307, 9 (93-140); 1320, 2 (92-110); 1324, 1 (84); 1327, 2 (59-114); 1342, 1 (128); 1359, 1 (118); 1361, 1 (116); 1364, 1 (110); 1365, 3 (92-112); 1374, 4 (78-117); 1380, 5 (35-59); 1383, 4 (96-105); 1388, 3 (94-123); 1392, 3 (94-102); 1393, 1 (93); 1397, 1 (98); 1432, 1 (134); 1448, 3 (90-135); 1454, 4 (62-114); 1456, 4 (70-106); 1462, 5 (106-138); 1468, 1 (123); 1470, 1 (126); 1480, 1 (81); 1507, 3 (106-127); 1510, 2 (109-126); 1513, 1 (113); 1516, 2 (125-152); 1518, 5 (110-140); 1522, 3 (91-126); 1525, 1 (120); 1528, 2 (89-119); 1590, 1 (147); 1606, 1 (122); 1607, 2 (107-150); 1608, 1 (129); 1615, 5 (71-119); 1623, 1 (119); 1634, 4 (104-136); 1636, 1 (96); 1637, 1 (122); 1641, 4 (85-117); 1645, 9 (107-139); 1648, 3 (108-112); 1649, 5 (90-140); 1653, 3 (118-137); 1658, 1 (113); 1661, 7 (73-134); 1666, 1 (37); 1671, 1 (133); 1676, 3 (92-140); 1678, 3 (98-110); 1679, 1 (111); 1685, 1 (135); 1686, 1 (112); 1689, 2 (88-100); 1692, 1 (98); 1731, 1 (48); 1734, 1 (44); 1755, 1 (144); 1777, 1 (144); 1825, 1 (38); 1829, 1 (40); 1830, 7 (79-92); 1835, 1 (50); 1836, 5 (46-59); 1839, 1 (70); 1842, 1 (125); 1977, 2 (60-69); 2111, 1 (125); 2174, 1 (109); 2177, 1 (110); 2183, 2 (126-130); 2187, 1 (102); 2189, 1 (118); 2191, 2 (60-133); 2210, 2 (88-128); 2216, 8 (85-105); 2217, 1 (132); 2226, 5 (41-61); 2232, 4 (44-87); 2234, 1 (?); 2237, 2 (118-143); 2241, 2 (106-116); 2242, 1 (119); 2244, 2 (110-130); 2245, 6 (62-140); 2247, 1

(127); 2253, 2 (88-114); 2260, 1 (133); 2262, 2 (114-125); 2263, 2 (87); 2268, 4 (63-136); 2283, 4 (32-50); 2286, 3 (53-60); 2287, 1 (113); 2289, 4 (62-80); 2290, 1 (80); 2291, 9 (67-111); 2292, 1 (116); 2293, 2 (114-130); 2294, 3 (59-99); 2300, 5 (54-131); 2301, 8 (52-146); 2299, 1 (140).

Eltanin (SOSC). 47, 1 (114); 54, 1 (105); 56, 2 (121-135); 59, 1 (130); 62, 1 (123); 77, 4 (97-127); 80, 4 (56-142); 144, 1 (93); 145, 2 (100-112); 147, 1 (103); 149, 2 (110-114); 162, 2 (76-123); 168, 1 (145); 302, 2 (104-142); 310, 1 (157); 311, 2 (112-143); 314, 1 (138); 321, 1 (149); 322, 4 (126-133); 328, 7 (33-89); 338, 1 (88); 340, 2 (48-59); 344, 4 (67-123); 352, 8 (90-128); 356, 2 (102-110); 363, 4 (53-94).

Discovery. 101, 1 (142); 107, 17 (58-104); 239, 1 (136); 407, 1 (86); 1775, 1 (54).

Lampanyctus sp. A

Eltanin (USC). 741, 8 (34-55); 1285, 22 (29-43); 1397, 1 (64); 1695, 1 (32); 1720, 12 (33-84); 1723, 21 (36-83); 1724, 24 (33-75); 1725, 1 (56); 1726, 23 (31-81); 1727, 2 (39-64); 1728, 5 (37-73); 1731, 7 (34-60); 1734, 10 (31-53); 1736, 25 (32-78); 1737, 13 (34-76); 1739, 5 (29-56); 1740, 7 (30-37); 1741, 2 (35-55); 1753, 113 (29-53); 1754, 1 (36); 1755, 69 (31-70); 1756, 3 (30-36); 1761, 15 (30-67); 1764, 8 (30-73); 1765, 6 (30-84); 1766, 29 (28-79); 1768, 3 (31-69); 1769, 2 (31-56); 1772, 9 (25-32); 1773, 7 (33-64); 1774, 7 (32-77); 1776, 3 (34-66); 1777, 32 (31-75); 1781, 31 (32-71); 1786, 8 (29-71); 1787, 11 (28-47); 1788, 1 (29); 1793, 17 (28-65); 1798, 10 (31-35); 1799, 20 (27-66); 1802, 29 (27-62); 1803, 7 (24-28); 1804, 6 (25-50); 1807, 14 (29-70); 1808, 2 (57-58); 1809, 43 (28-52); 1810, 16 (27-56); 1811, 3 (61-78); 1812, 47 (31-73); 1825, 1 (35).

Eltanin (SOSC). 85, 1 (70); 92, 7 (33-73); 305, 2 (45-58); 307, 31 (35-71); 311, 8 (51-67); 314, 1 (74); 315, 27 (35-61); 316, 6 (29-56); 317, 23 (30-75); 320, 4 (31-35); 321, 43 (33-65); 322, 1 (73); 323, 1 (41); 328, 1 (77); 335, 1 (66); 336, 1 (68); 340, 1 (58); 341, 2 (58-80); 345, 2 (62-75).

Discovery. 81, 2 (36-50); 86, 3 (55-57); 87, 1 (38); 252, 1 (40); 1805, 2 (41-62).

Lampanyctus pusillus

Eltanin (USC). 1287, 2 (24-27); 1720, 1 (25); 1724, 1 (25); 1736, 1 (28); 1739, 2 (22-32); 1758, 3 (23-28); 1766, 3 (23-27); 1772, 2 (21-23); 1787, 1 (23); 1788, 15 (21-34); 1794, 1 (22); 1807, 2 (25-30); 1809, 1 (32); 1810, 2 (26-33); 1811, 1 (30); 1820, 1 (33); 1825, 1 (34); 2222, 4 (26-30); 2223, 1 (28); 2224, 2 (21-22); 2226, 1 (33); 2227, 1 (30); 2228, 1 (31); 2271, 259 (20-34); 2273, 21 (22-32); 2278, 39 (22-35); 2279, 11 (22-36); 2280, 10 (26-32); 2281, 7 (21-27); 2283, 2 (22-27).

Eltanin (SOSC). 305, 1 (32); 307, 11 (15-31); 308, 1 (31); 315, 1 (24).

Discovery. 81, 1 (28); 85, 1 (28); 86, 1 (32); 89, 1 (25); 100, 2 (18-28); 100c, 2 (32-34); 101, 1 (36); 257, 2 (30-32); 402, 1 (25); 413, 1 (22); 419, 1 (21); 435, 1 (26); 673, 3 (22-28); 873, 4 (19-31); 877, 1 (27); 1371, 1 (16); 1602, 3 (21-24); 1608, 1 (32); 1736, 3 (28-32); 1749, 1 (23); 1758, 1 (30); 1759, 1 (29); 1768, 1 (36); 1771, 1 (19); 1806, 1 (24); 2067, 1 (17); 2152, 4

(17-33); 2153, 2 (19-21); 2729, 1 (18); 2730, 1 (32); 2731, 2 (23); 3203, 3 (21-29).

Lampanyctus australis

Eltanin (USC). 97, 17 (25-99); 99, 2 (62-76); 165, 293 (19-104); 169, 103 (23-103); 175, 12 (27-96); 213, 39 (18-111); 215, 56 (26-100); 326, 5 (34-88); 563, 1 (27); 670, 1 (79); 741, 24 (25-98); 957, 1 (84); 1185, 1 (39); 1286, 12 (19-105); 1287, 15 (28-110); 1288, 8 (35-42); 1397, 2 (33-87); 1401, 6 (39-94); 1402, 14 (24-92); 1405, 37 (23-97); 1409, 6 (66-88); 1432, 1 (91); 1525, 8 (18-93); 1695, 9 (27-93); 1696, 8 (25-98); 1697, 1 (67); 1700, 2 (69-99); 1701, 68 (22-48); 1704, 47 (31-98); 1706, 10 (32-93); 1710, 11 (30-95); 1723, 2 (19-37); 1726, 1 (66); 1728, 2 (64-88); 1731, 1 (31); 1740, 1 (34); 1755, 1 (77); 1817, 21 (16-22); 1820, 7 (13-99); 1821, 6 (44-103); 1823, 11 (20-100); 1824, 4 (60-98); 1825, 6 (18-97); 1830, 24 (41-95); 1832, 11 (58-97); 1834, 2 (81-89); 1835, 5 (64-93); 1838, 36 (49-96); 1839, 4 (53-83); 1841, 2 (51-82); 1985, 6 (19-95); 2217, 4 (72-85); 2221, 7 (67-90); 2222, 11 (39-50); 2223, 8 (33-107); 2224, 5 (22-96); 2226, 19 (30-97); 2227, 7 (33-84); 2228, 1 (42); 2231, 11 (34-99); 2232, 48 (31-101); 2233, 7 (41-88); 2250, 1 (88); 2252, 5 (67-83); 2269, 3 (33-99); 2270, 8 (29-72); 2271, 4 (32-40); 2278, 5 (26-29); 2279, 4 (25-46); 2280, 3 (23-54); 2281, 37 (30-78); 2283, 7 (30-37); 2285, 3 (80-90); 2302, 27 (70-99).
Eltanin (SOSC). 88, 4 (85-99); 92, 1 (87); 302, 2 (30-32); 307, 1 (28); 315, 1 (21); 321, 2 (30-108); 322, 11 (36-102); 323, 1 (57).
Discovery. 83, 1 (63); 86, 1 (88); 101, 5 (50-90); 239, 7 (75-99); 241, 1 (39); 257, 2 (67-72); 407, 12 (81-104); 413, 4 (18-31); 420, 1 (49); 440, 2 (93-96); 441, 1 (56); 672, 1 (89); 673, 1 (21); 717, 3 (58-65); 718, 2 (54-65); 1554, 1 (78); 1571, 1 (98); 1602, 2 (18-27); 1749, 1 (30); 1758, 1 (36); 1768, 1 (28); 1770, 1 (79); 1871, 1 (27); 2035, 1 (88).
BANZARE. 111, 1 (79).

Lampanyctus alatus

Eltanin (USC). 2278, 2 (20-57); 2281, 1 (24).
Discovery. 83, 1 (40); 281, 2 (27-30); 284, 2 (25-42); 285, 4 (26-53); 286, 8 (20-53); 288, 1 (40); 294, 1 (42); 296, 6 (20-43); 297, 4 (30-37); 404, 2 (35-37); 407, 4 (34-43); 439, 1 (19); 440, 2 (27-44); 692, 1 (44); 695, 1 (44); 701, 2 (28-33); 702, 2 (21-57); 704, 1 (45); 875, 1 (58); 1371, 1 (41); 1372, 1 (33); 1582, 3 (35-39); 1598, 3 (29-45); 1600, 2 (32-46); 1736, 2 (19-49); 1764, 1 (54); 1768, 1 (38); 2036, 1 (33); 2048, 1 (55); 2055, 1 (20); 2057, 3 (47-55); 2059, 4 (26-35); 2061, 1 (51); 2063, 1 (45); 2065, 1 (41); 2066, 1 (40); 2067, 2 (22-29); 2149, 1 (40); 2153, 1 (42); 2729, 2 (22-35); 2797, 5 (18-51).

Lampanyctus macdonaldi

Eltanin (USC). 97, 2 (104-109); 99, 7 (82-113); 110, 1 (70); 123, 2 (73-74); 165, 3 (43-69); 213, 1 (29); 215, 9 (90-129); 306, 1 (84); 354, 1 (87); 359, 1 (76); 360, 1 (110); 381, 1 (62); 741, 1 (40); 767, 1 (73); 778, 1 (108); 846, 2 (71-120); 849, 7 (51-117); 852, 1

(92); 868, 6 (54-100); 874, 1 (69); 877, 1 (125); 882, 1 (63); 889, 1 (105); 890, 1 (38); 900, 1 (46); 904, 1 (112); 1170, 2 (60-68); 1287, 1 (121); 1290, 1 (61); 1294, 1 (75); 1302, 1 (46); 1303, 1 (80); 1380, 2 (34-44); 1396, 1 (106); 1397, 3 (35-123); 1401, 1 (116); 1402, 1 (123); 1454, 1 (51); 1456, 1 (90); 1522, 1 (70); 1525, 1 (85); 1586, 1 (65); 1607, 1 (102); 1608, 1 (90); 1615, 1 (55); 1645, 2 (80-87); 1696, 3 (25-27); 1697, 9 (73-115); 1700, 15 (40-125); 1710, 2 (38-52); 1724, 1 (90); 1727, 1 (110); 1731, 1 (43); 1740, 1 (86); 1755, 1 (35); 1821, 1 (104); 1830, 36 (35-87); 1834, 55 (35-107); 1835, 10 (57-116); 1836, 1 (58); 1839, 8 (25-103); 1842, 13 (35-115); 1985, 8 (51-106); 2216, 5 (50-70); 2221, 24 (28-112); 2224, 5 (32-60); 2226, 11 (28-65); 2227, 1 (60); 2231, 1 (49); 2232, 11 (29-51); 2233, 4 (27-45); 2247, 2 (79-110); 2253, 10 (39-123); 2269, 8 (28-69); 2283, 6 (28-46); 2285, 2 (114-115); 2286, 5 (40-52); 2287, 1 (108); 2290,, 1 (115).
Eltanin (SOSC). 80, 3 (68-75); 168, 1 (58); 307, 1 (44); 323, 5 (107-120); 328, 2 (45-48); 344, 6 (62-88).
Discovery. 101, 2 (83-111).

Lampanyctus intricarius

Eltanin (USC). 99, 2 (102-113); 1168, 2 (125-177); 1286, 1 (?90); 1402, 1 (28); 1409, 2 (73-79); 1695, 1 (45); 1723, 1 (36); 1725, 1 (36); 1726, 3 (33-40); 1736, 2 (34-36); 1738, 2 (34-36); 1740, 1 (40); 1755, 1 (71); 1761, 2 (33-39); 1765, 1 (36); 1766, 4 (33-43); 1772, 5 (33-41); 1773, 1 (37); 1774, 1 (36); 1781, 2 (35-48); 1788, 3 (28-32); 1793, 1 (35); 1810, 1 (30); 1811, 2 (34-44); 1839, 3 (139-150); 2221, 1 (138); 2222, 2 (41-45); 2269, 1 (127); 2281, 2 (50-53); 2286, 1 (139); 2287, 4 (42-96); 2288, 1 (112); 2294, 1 (136); 2301, 1 (101).
Eltanin (SOSC). 303, 12 (67-135); 314, 1 (35); 315, 5 (32-67); 316, 2 (36-40); 317, 8 (37-44); 321, 2 (40-72); 338, 1 (102).
Discovery. 87, 2 (48-51); 107, 1 (108); 250, 1 (42); 841, 1 (34); 1734, 1 (32).
British Museum of Natural History. 1902-11-26-2 (from stomach of Melanocetus johnsoni, 31°N, 37°W), 1 (179).

Lampanyctus lepidolychnus

Eltanin (USC). 1696, 1 (105); 1700, 1 (114); 2227, 1 (60); 2285, 1 (69); 2288, 1 (78); 2302, 2 (93-99).
Discovery. 76, 2 (62-96); 100a, 2 (58-63); 440, 1 (34); 1372, 1 (30); 1755, 1 (89); 1758, 1 (79); 1766, 1 (88); 1767, 1 (82); 1799, 1 (74); 2035, 1 (42).

Lampanyctus iselinoides

Eltanin (USC). 165, 5 (72-84); 169, 6 (28-90); 175, 4 (24-89); 213, 17 (24-84); 215, 4 (27-84); 326, 44 (19-89); 741, 6 (23-81); 1286, 38 (22-91); 1287, 30 (25-86); 1288, 6 (80-94).
Eltanin (SOSC). 88, 15 (44-87); 301, 1 (42); 302, 41 (20-94); 303, 488 (25-97); 322, 1 (29).
William Scoresby. 599, 8 (24-55); 600, 3 (19-64); 600, 20 (20-47); 600, 15 (20-45); 601, 17 (26-63); 604, 1 (33); 605, 3 (20-28); 740, 1 (77); 741, 4 (24-64).

Lampanyctus sp. B

Eltanin (USC). 1401, 1 (68).

Lampanyctus sp. C

Eltanin (USC). 766, 1 (21); 1768, 1 (48); 1769, 1 (24); 1786, 6 (23-48); 1787, 4 (22-29); 1788, 24 (22-47); 1807, 3 (23-24); 1809, 1 (24); 1810, 5 (19-20).

Lampanyctus sp. D

Eltanin (USC). 165, 1 (28); 1285, 24 (27-54); 1287, 4 (32-53); 1288, 5 (30-46); 1525, 1 (44); 1701, 3 (21-48); 1704, 1 (56); 1710, 3 (20-26); 1720, 10 (22-53); 1723, 12 (21-51); 1724, 20 (21-62); 1726, 21 (23-54); 1727, 1 (22); 1728, 1 (44); 1731, 1 (24); 1734, 5 (21-48); 1736, 4 (22-24); 1737, 2 (22-48); 1738, 44 (22-52); 1739, 4 (21-25); 1740, 9 (20-29); 1741, 3 (21-27); 1753, 39 (23-67); 1755, 2 (36-54); 1756, 3 (28-47); 1758, 94 (22-34); 1761, 1 (28); 1764, 1 (25); 1765, 4 (21-52); 1766, 1 (24); 1772, 42 (22-63); 1773, 5 (21-54); 1774, 3 (24-62); 1776, 4 (25-54); 1777, 5 (22-27); 1781, 11 (22-58); 1787, 1 (26); 1788, 2 (23-24); 1793, 10 (21-28); 1794, 9 (26-49); 1798, 26 (22-34); 1799, 16 (22-50); 1802, 28 (21-44); 1803, 95 (23-57); 1804, 12 (21-55); 1811, 1 (25); 1812, 1 (21); 1823, 2 (43-44); 1830, 1 (42); 2222, 7 (22-26); 2271, 11 (25-40); 2278, 2 (27-28); 2281, 6 (23-28).
Eltanin (SOSC). 92, 2 (21-53); 306, 13 (21-34); 307, 4 (23-31); 315, 29 (25-49); 316, 5 (28-31); 317, 6 (28-33); 321, 12 (23-48); 322, 1 (46); 323, 4 (22-39); 338, 1 (41).

Lobianchia dofleini

Eltanin (USC). 1525, 4 (16-36); 1719, 2 (28-32); 1720, 1 (29); 1723, 1 (29); 1726, 2 (24-32); 1734, 1 (29); 1736, 2 (27-35); 1738, 1 (27); 1740, 1 (33); 1753, 1 (28); 1755, 3 (32-35); 1758, 6 (25-30); 1765, 1 (25); 1772, 10 (25-32); 1773, 1 (26); 1774, 4 (30-32); 1776, 1 (34); 1777, 1(29); 1781, 1 (32); 1787, 1 (24); 1788, 1 (26); 1793, 1 (28); 1794, 8 (28-35); 1799, 5 (28-31); 1802, 6 (27-46); 1803, 10 (27-41); 1804, 1 (28); 1811, 1 (40); 1817, 2 (15-41); 1820, 2 (15-32); 1823, 1 (37); 1825, 1 (36); 1984, 1 (25); 2222, 1 (30); 2223, 1 (28); 2271, 1 (31); 2273, 8 (25-38); 2278, 12 (25-39); 2279, 3 (28-29); 2280, 3 (29-32); 2281, 1 (29).
Eltanin (SOSC). 307, 1 (40); 315, 3 (32-35); 317, 2 (36-38).
Discovery. 250, 2 (31-40); 251, 1 (46); 254, 2 (34-35); 285, 2 (33-38); 286, 1 (44); 872, 1 (29); 1568, 1 (27); 1602, 2 (18-19); 2067, 1 (17).
William Scoresby. 607, 2 (30-32).

Diaphus ostenfeldi

Eltanin (USC). 165, 2 (76-84); 169, 2 (57-58); 213, 1 (57); 1100, 3 (79-90); 1107, 1 (83); 1167, 1 (84); 1405, 3 (75-88); 1409, 5 (74-91); 1525, 1 (55); 1720, 1 (30); 1730, 1 (84); 1738, 3 (30-31); 1756, 5 (30-35); 1772, 6 (29-32); 1781, 1 (32); 1786, 1 (86); 1794, 5 (31-78); 1798, 4 (32-73); 1799, 6 (30-70); 1803, 16 (27-86); 1807, 3 (87-110); 1810, 1 (74); 1811, 5 (31-56); 1817,

1 (21); 1820, 1 (109); 2221, 1 (84); 2250, 1 (77); 2270, 4 (60-70); 2271, 2 (35-118); 2273, 19 (26-38); 2278, 1 (34); 2279, 1 (34); 2281, 1 (66).
Eltanin (SOSC). 306, 2 (34-47); 315, 4 (56-92); 317, 1 (51); 327, 1 (55).
Discovery. 2492, 1 (82); 2582, 1 (89).
BANZARE. 71, 1 (76).

Diaphus effulgens

Eltanin (USC). 1736, 1 (77); 1764, 1 (82); 1787, 1 (72).
Discovery. 1375, 1 (50).

Diaphus danae

Eltanin (USC). 1405, 2 (67-68); 1821, 3 (25-32); 1823, 4 (23-31); 1824, 1 (30); 2222, 1 (33); 2223, 1 (28).

Diaphus mollis

Eltanin (USC). 2223, 2 (44-62).

Diaphus termophilus

Eltanin (USC). 1821, 1 (36).

Diaphus parri

Eltanin (USC). 1402, 1 (36); 1720, 1 (47); 1723, 1 (44); 1726, 3 (43-46); 1736, 2 (31-46); 1737, 1 (42); 1741, 2 (20-45); 1758, 1 (35); 1772, 3 (26-30); 1785, 2 (22-54); 1788, 5 (22-27); 1794, 1 (40); 1808, 4 (20-22); 1817, 2 (32-33); 1823, 12 (29-36); 2222, 11 (13-37); 2224, 1 (34); 2273, 4 (21-28); 2278, 4 (23-42); 2279, 2 (18-36); 2280, 1 (34).
Discovery. 433, 1 (31); 1771, 1 (15).

Diaphus sp. A

Eltanin (USC). 97, 1 (50); 169, 11 (23-75); 213, 1 (36); 326, 1 (34); 741, 15 (21-47); 1286, 1 (35); 1287, 3 (32-33); 1288, 5 (32-56); 1402, 4 (40-60); 1409, 4 (58-67); 1427, 1 (50); 1694, 7 (36-49); 1695, 1 (68); 1697, 2 (44-47); 1706, 4 (16-47); 1710, 2 (44-48); 1724, 1 (67); 1726, 3 (17-20); 1728, 1 (40); 1734, 1 (19); 1738, 1 (18); 1739, 1 (20); 1740, 2 (17-21); 1758, 1 (21); 1761, 1 (19); 1772, 5 (16-21); 1777, 1 (54); 1797, 1 (20); 1816, 1 (33); 1817, 1 (72); 1820, 2 (33-35); 1821, 1 (71); 1823, 3 (37-41); 1824, 1 (59); 1832, 4 (34-40); 1839, 1 (61); 1842, 1 (61); 1984, 2 (42-43); 2217, 1 (63); 2223, 3 (18-75); 2224, 1 (72); 2226, 1 (26); 2232, 2 (51-76); 2270, 1 (30); 2273, 1 (170, 2285, 1 (50); 2288, 1 (50); 2302, 2 (35-60).
Eltanin (SOSC). 88, 1 (51); 303, 1 (78); 307, 2 (22-27); 315, 1 (72); 316, 1 (66); 322, 2 (64-67); 327, 1 (57).
Discovery. 402, 1 (18); 1188, 1 (36); 1280, 1 (48); 1612, 1 (35); 2154, 1 (46); 2883, 1 (60).
William Scoresby. 588, 1 (48).

Diaphus sp. B

Eltanin (USC). 2252, 1 (51); 2273, 2 (16-17); 2278, 5 (15-20); 2279, 1 (17); 2280, 1 (22); 283, 1 (50).

Hintonia candens

Eltanin (USC). 97, 2 (43-50); 165, 5 (50-68);
169, 3 (53-60); 1288, 1 (105); 1405, 2 (40-44);
1409, 1 (24); 1696, 2 (52-55); 1720, 2 (47-54);
1724, 1 (51); 1726, 4 (43-79); 1727, 3 (35-45);
1737, 1 (50); 1755, 1 (45); 1756, 3 (43-49);
1777, 1 (87); 1781, 1 (59); 1803, 1 (53); 1824,
2 (22-24); 1829, 1 (25); 1830, 2 (22-29); 1838, 4
(23-84); 1839, 5 (23-25); 1840, 1 (24); 1842, 2
(23); 2221, 1 (53); 2226, 1 (100); 2232, 1 (59);
2286, 1 (58); 2301, 1 (83); 2302, 3 (70-95).
Eltanin (SOSC). 322, 1 (110); 328, 1 (115);
348, 1 (79).
Discovery. 1774, 1 (83); 2531, 1 (77).
William Scoresby. 588, 1 (50).

Lampanyctodes hectoris

Eltanin (USC). 326, 27 (14-49); 336, 1 (18);
1286, 3 (16-17); 1398, 6 (45-51); 1400, 18
(28-59); 1401, 4 (20-35); 1427, 3 (52-56); 1428,
2 (55-56); 1432, 1 (44); 1713, 2 (39-51); 1824, 1
(28); 1840, 31 (22-24).
Discovery. 99d, 1 (56); 100c, 2 (15-16); 1281,
3 (32-37); 2714, 1 (21).
William Scoresby. 745, 3 (41-49).
British   Museum   of   Natural   History.
1876-2-12-18 (Cook Strait, New Zealand), 1 (53);
1902-5-28-17 (Table Bay, South Africa), 1 (43);
1927-12-6-4-6 (off Cape point, South Africa), 3
(38-45);   1947-12-27-1-2   (Table   Bay,   South
Africa), 2 (38-47); 1947-12-24-3-5 (off Cape
Point, South Africa), 3 (46-48).

Scopelopsis multipunctatus

Eltanin (USC).   1766,   5   (44-46);   2278,   4
(24-61); 2280, 2 (20-22).
Discovery. 101, 1 (35); 413, 3 (19-22); 433, 2
(18-21); 434, 1 (19); 440, 1 (30); 442, 1 (29);
712, 1 (25); 713, 1 (20); 1763, 2 (48-54); 1766,
1 (52); 2073, 1 (27).
British   Museum   of   Natural   History.
1912-3-1-115 (near Agulhas Bank, South Africa), 1
(54).

Notoscopelus resplendens

Eltanin (USC). 1788, 1 (59).
Discovery. 168, 1 (58); 287, 1 (29); 288, 1
(52); 293, 1 (49); 448, 1 (55); 1584, 1 (63);
2730, 1 (41).
British   Museum   of   Natural   History.
1926-6-30-4-5 (Lord Howe island), 2 (65-67).

Lampichthys procerus

Eltanin (USC). 667, 2 (57-58); 741, 1 (55);
1286, 6 (20-86); 1287, 4 (20-61); 1288, 5
(59-77); 1401, 7 (20-67); 1402, 2 (23-33); 1405,
2 (74-75); 1695, 1 (43); 1700, 3 (53-59); 1704, 3
(52-75); 1706, 1 (82); 1720, 2 (57-74); 1723, 1
(60); 1724, 1 (78); 1726, 3 (54-70); 1727, 1
(58); 1728, 3 (57-74); 1758, 1 (58); 1774, 1
(59); 1793, 4 (54-81); 1794, 1 (62); 1804, 1
(62); 1817, 11 (26-35); 1820, 6 (20-28); 1821, 2
(30-31); 1823, 2 (30-36); 1824, 6 (21-81); 1825,
2 (27-77); 1830, 19 (24-73); 1831, 356 (19-75);
1832, 15 (27-33); 1834, 1 (34); 1835, 2 (22-28);
1838, 4 (23-67); 1984, 1 (50); 1985, 2 (44-45);

2232, 4 (56-67); 2278, 2 (79-80); 2281, 1 (58);
2302, 2 (62-69).
Eltanin (SOSC). 88, 2 (45-49); 307, 2 (20);
311, 2 (21-64); 314, 5 (19-22); 315, 8 (19-23);
317, 4 (20-22); 321, 2 (21).
Discovery. 717, 1 (76); 1188, 1 (22); 1281, 1
(78).

Gymnoscopelus braueri

Eltanin (USC). 97, 1 (99); 99, 24 (30-52); 109,
7 (27-100); 123, 1 (109); 133, 1 (75); 134, 1
(45); 137, 2 (36-66); 138, 1 (91); 141, 2
(38-42); 143, 1 (48); 148, 5 (92-103); 149, 2
(29-99); 154, 3 (85-92); 232, 1 (95); 235, 2
(39-106); 236, 54 (30-55); 246, 1 (38); 247, 2
(37-73); 248, 1 (53); 252, 1 (52); 253, 6
(36-119); 259, 1 (64); 262, 1 (53); 279, 1 (97);
281, 1 (80); 302, 6 (72-107); 304, 3 (53-62);
306, 7 (31-96); 310, 1 (30); 318, 4 (92-107);
348, 5 (35-101); 359, 1 (105); 360, 3 (35-96);
364, 2 (44-49); 368, 12 (39-85); 375, 1 (32);
378, 1 (34); 382, 1 (36); 392, 3 (33-95); 396, 17
(53-113); 414, 3 (64-106); 422, 7 (80-125); 431,
3 (85-107); 448, 1 (104); 483, 4 (?); 508, 2
(82-91); 550, 1 (86); 563, 20 (21-25); 567, 1
(48); 570, 1 (117); 571, 5 (43-111); 572, 1 (62);
575, 1 (96); 580, 1 (48); 582, 2 (72-76); 592, 1
(93); 593, 9 (40-109); 597, 2 (43-74); 601, 1
(112); 605, 1 (57); 627, 1 (103); 632, 2
(105-108); 634, 2 (49-70); 640, 1 (48); 642, 2
(67-73); 643, 5 (62-102); 654, 2 (62-90); 667, 2
(34-47); 670, 2 (33); 683, 2 (63-65); 687, 1
(60); 696, 1 (60); 697, 5 (58-99); 702, 4
(54-99); 703, 2 (70-83); 714, 1 (65); 718, 2
(37-59); 719, 19 (46-89); 730, 1 (49); 737, 1
(63); 767, 2 (35-102); 771, 2 (38-114); 775, 2
(32-41); 781, 4 (30-47); 782, 3 (30-89); 785, 4
(30-45); 788, 14 (34-61); 793, 2 (90-118); 802, 2
(83-113); 811, 4 (48-99); 812, 2 (58-100); 832, 2
(42-51); 835, 1 (37); 836, 15 (30-85); 839, 1
(33); 846, 2 (40-114); 849, 1 (35); 850, 2
(31-33); 852, 2 (33-39); 854, 3 (50-89); 855, 1
(59); 858, 2 (63-95); 859, 5 (72-109); 864, 3
(97-123); 866, 6 (39-96); 867, 4 (33-98); 868, 6
(28-49); 874, 1 (34); 882, 2 (35-36); 883, 2
(32-35); 886, 4 (32-43); 888, 4 (29-93); 889, 4
(36-63); 890, 8 (35-114); 891, 8 (35-92); 895, 5
(35-52); 898, 2 (37-92); 900, 6 (35-91); 901, 3
(36-55); 903, 3 (32-52); 904, 2 (32-34); 906, 3
(33-59); 911, 2 (62-63); 912, 2 (49-81); 914, 3
(34-96); 917, 4 (59-91); 918, 2 (59-83); 919, 1
(89); 920, 2 (91-123); 929, 12 (81-104); 930, 1
(98); 935, 11 (78-126); 936, 1 (89); 940, 1 (84);
941, 1 (95); 943, 9 (62-96); 944, 1 (79); 947, 1
(91); 949, 2 (41-91); 950, 3 (61-106); 952, 2
(57-96); 953, 4 (34-89); 957, 2 (27-41); 998, 3
(95-108); 1006, 3 (102-113); 1014, 1 (97); 1015,
1 (105); 1019, 1 (96); 1020, 4 (93-102); 1022, 2
(92-102); 1023, 6 (93-109); 1026, 23 (80-116);
1029, 11 (74-122); 1030, 2 (72-93); 1036, 4
(77-108); 1044, 6 (94-119); 1050, 5 (97-106);
1051, 1 (115); 1057, 5 (84-116); 1064, 6
(92-106); 1065, 3 (74-108); 1071, 7 (70-104);
1077, 2 (89-96); 1100, 3 (26-28); 1106, 1 (89);
1107, 4 (27-28); 1113, 1 (26); 1121, 4 (38-53);
1132, 1 (75); 1133, 1 (81); 1137, 1 (89); 1141, 2
(79-87); 1142, 17 (75-104); 1162, 2 (76-93);
1163, 5 (81-98); 1167, 1 (30); 1170, 2 (28-115);
1186, 2 (31-32); 1187, 1 (92); 1195, 2 (98-109);
1196, 6 (29-45); 1203, 1 (42); 1204, 24 (29-109);

1206, 9 (27-83); 1210, 5 (30-80); 1213, 3 (87-99); 1214, 4 (36-91); 1220, 1 (41); 1236, 3 (66-99); 1241, 6 (79-102); 1245, 3 (65-98); 1262, 4 (29-91); 1269, 7 (29-47); 1290, 9 (27-95); 1294, 9 (28-75); 1299, 18 (28-98); 1302, 3 (28-89); 1303, 6 (32-94); 1304, 43 (30-46); 1307, 3 (29-91); 1315, 2 (29-72); 1316, 25 (29-57); 1320, 1 (29); 1323, 2 (33-108); 1325, 10 (30-104) 1327, 4 (30-37); 1328, 1 (28); 1332, 1 (31) 1333, 23 (31-51); 1336, 49 (30-44); 1337, 9 (29-33); 1342, 2 (32-34); 1348, 3 (30-44); 1350, 5 (30-35); 1355, 20 (35-119); 1361, 1 (34); 1363, 3 (42-95); 1364, 5 (33-110); 1365, 4 (32-56); 1374, 17 (32-56); 1379, 2 (29-31); 1383, 1 (34); 1384, 1 (48); 1389, 2 (28-33); 1392, 5 (55-113); 1393, 1 (59); 1439, 9 (26-30); 1448, 3 (27-28); 1454, 2 (30-31); 1456, 1 (30); 1462, 3 (41-88); 1463, 2 (36-100); 1468, 1 (53); 1470, 2 (59-94); 1473, 2 (41-99); 1475, 3 (30-106); 1481, 2 (31-32); 1485, 2 (49-75); 1488, 2 (30-74); 1507, 1 (45); 1510, 3 (57-108); 1513, 3 (35-56); 1522, 5 (95-109); 1525, 2 (35-36); 1528, 2 (41-52); 1538, 1 (62); 1543, 4 (59-78); 1552, 4 (88-113); 1574, 1 (48); 1586, 2 (46-95); 1606, 7 (26-27); 1607, 23 (24-28); 1608, 1 (26); 1609, 3 (36-82); 1615,, 17 (28-115); 1616, 1 (88); 1622, 3 (38-52); 1623, 1 (83); 1627, 2 (36-37); 1633, 3 (51-98); 1634, 5 (36-100); 1635, 5 (36-100); 1636, 1 (38); 1637, 3 (50-60); 1641, 3 (37-87); 1645, 7 (38-110); 1648, 5 (39-50); 1649, 1 (36); 1653, 5 (26-55); 1658, 2 (28-39); 1661, 1 (83); 1665, 2 (83-109); 1666, 1 (50); 1671, 1 (65); 1672, 30 (33-58); 1676, 9 (49-104); 1677, 12 (40-110); 1679, 2 (26-120); 1683, 3 (36-83); 1684, 16 (27-105); 1685, 4 (23-29); 1686, 24 (27-29); 1687, 6 (25-28); 1689, 11 (23-28); 1692, 3 (37-96); 1855, 2 (32-62); 1865, 1 (85); 1947, 1 (115); 1955, 2 (58-96); 1959, 4 (76-105); 1970, 2 (37-39); 1976, 6 (32-96); 1977, 1 (70); 1992, 3 (72-94); 2111, 1 (112); 2139, 1 (34); 2174, 4 (42-55); 2177, 1 (50); 2179, 11 (40-71); 2183, 14 (37-103); 2187, 11 (26-54); 2191, 1 (29); 2204, 2 (30-31); 2205, 7 (27-31); 2210, 5 (51-87); 2211, 4 (61-98); 2212, 3 (75-92); 2213, 16 (31-115); 2216, 1 (29); 2218, 15 (26-31); 2235, 2 (40-41); 2236, 11 (43-93); 2237, 5 (41-88); 2238a, 1 (78); 2238b, 5 (50-87); 2239, 7 (41-80); 2240, 3 (59-81); 2241, 19 (50-117); 2242, 1 (44); 2243, 3 (41-83); 2244, 2 (41-67); 2245, 5 (30-121); 2246, 6 (30-106); 2254,. 2 (32-40); 2260, 1 (44); 2261, 4 (44-76); 2262, 6 (43-80); 2263, 6 (33-81); 2265, 2 (74-85); 2291, 2 (52-107); 2292, 5 (47-114); 2293, 3 (34-85); 2294, 8 (32-110); 2295, 3 (40-75); 2296, 13 (42-117); 2297, 2 (58-88); 2298, 11 (31-97); 2299, 3 (30-68); 2300, 7 (29-31).

Eltanin (SOSC). 11, 15 (34-52); 26, 5 (51-104); 28, 1 (55); 41, 4 (85-89); 47, 4 (86-98); 49, 8 (61-108); 54, 2 (50-67); 59, 2 (102-115); 63, 1 (78); 69, 2 (35-57); 74, 25 (37-79); 77, 5 (35-75); 80, 2 (23-53); 82, 1 (30); 85, 1 (27); 118, 3 (57-83); 124, 4 (76-94); 125, 1 (80); 139, 5 (49-119); 140, 1 (102); 144, 2 (34-57); 145, 4 (45-103); 147, 5 (32-66); 151, 5 (30-44); 159, 1 (33); 349, 3 (29-31); 352, 4 (30-92); 354, 4 (31-34); 356, 12 (30-93); 357, 4 (30-48); 358, 18 (30-80); 362, 13 (30-117).

Eltanin cruise 33 (ANARE). 33-5, 1 (108); 33-6, 2 (85-97); 33-7, 1 (95); 33-11, 1 (79); 33-12, 1 (104); 33-13, 1 (95); 33-15, 1 (100).

Discovery. 62, 1 (29); 65, 2 (47-62); 66, 3

(33-45); 239, 1 (31); 321, 1 (62); 344, 2 (63-69); 348, 1 (43); 357, 1 (92); 481, 2 (53-54); 493, 1 (59); 522, 1 (73); 530, 2 (33-68); 555, 1 (96); 668, 2 (25-29); 716, 1 (36); 718, 2 (48-57); 765,1 (?75); 778, 1 (?110); 780, 5 (?80-100); 799, 1 (96); 824, 1 (109); 825, 1 (61); 831, 2 (86-89); 853, 1 (70); 878, 1 (26); 1017, 1 (68); 1029, 1 (118); 1031, 2 (74-82); 1033, 1 (72); 1050, 1 (92); 1061, 3 (57-73); 1138, 2 (76-99); 1144, 2 (80-99); 1148, 2 (79-95); 1212, 5 (69-76); 1214, 1 (106); 1233, 1 (44); 1282, 1 (90); 1295, 1 (?115); 1309, 1 (?105); 1314, 2 (56-67); 1316, 1 (107); 1331, 2 (79-105); 1360, 1 (83); 1362, 1 (82); 1421, 1 (33); 1429, 1 (116); 1492, 4 (78-79); 1550, 3 (55-75); 1551, 3 (61-63); 1559, 18 (44-108); 1617, 2 (35-116); 1677, 1 (91); 1694, 1 (77); 1705, 1 (89); 1711, 1 (100); 1723, 1 (82); 1725, 2 (93-107); 1745, 1 (44); 1784, 1 (81); 1812, 1 (37); 1846, 1 (?65); 1858, 1 (90); 1918, 2 (50-96); 1947, 2 (?75-80); 1994, 1 (106); 2020, 1 (60); 2033, 1 (30); 2106, 2 (64-75); 2128, 1 (110); 2139, 1 (37); 2141, 1 (59); 2162, 1 (107); 2162, 1 (107); 2224, 1 (?45); 2307, 2 (50-66); 2308, 1 (108); 2567, 2 (86-92); 2610, 1 (76); 2612, 1 (65); 2614, 1 (80); 2616, 2 (76-91); 2618, 1 (108); 2704, 1 (81); 2809, 1 (40); 2825, 1 (27).

William Scoresby. 351, 3 (?40).

Gymnoscopelus opisthopterus

Eltanin (USC). 99, 5 (35-38); 123, 1 (126); 125, 2 (35-39); 132, 9 (44-158); 133, 2 (131-147); 137, 9 (51-153); 142, 1 (127); 154, 5 (41-126); 235, 1 (135); 247, 1 (151); 248, 3 (128-145); 253, 1 (37); 259, 2 (121-128); 262, 2 (137-149); 265, 1 (131); 274, 11 (104-151); 275, 2 (127-129); 279, 3 (59-136); 282, 2 (127-131); 285, 3 (60-140); 292, 2 (53-81); 297, 1 (86); 310, 1 (32); 364, 1 (142); 368, 2 (40-100); 381, 1 (127); 382, 1 (39); 383, 4 (47-140); 388, 1 (40); 396, 3 (54-147); 397, 2 (135-152); 449, 1 (44); 510, 1 (26); 532, 1 (72); 563, 10 (25-27); 567, 3 (52-56); 580, 1 (53); 581, 2 (51-75); 588, 1 (56); 597, 4 (61-155); 611, 2 (63-138); 625, 1 (91); 626, 2 (66-123); 634, 3 (53-102); 635, 1 (134); 643, 2 (107-150); 667, 1 (44); 670, 4 (48-67); 683, 5 (92-145); 687, 2 (92-150); 691, 1 (144); 692, 3 (54-96); 703, 3 (76-136); 714, 1 (122); 718, 2 (62-80); 729, 2 (60-65); 737, 1 (159); 738, 1 (35); 771, 1 (40); 782, 2 (37-43); 792, 4 (54-131); 796, 2 (51-57); 802, 2 (49-135); 832, 1 (136); 839, 1 (38); 847, 2 (50-141); 849, 2 (37-127); 850, 3 (46-136); 854, 1 (96); 855, 3 (53-126); 858, 3 (44-132); 864, 3 (37-157); 867, 1 (157); 874, 1 (139); 882, 1 (55), 895, 3 (42-81); 903, 2 (44-64); 914, 1 (84); 919, 3 (65-150); 933, 1 (111); 936, 2 (132-154); 940, 1 (51); 941, 1 (89); 944, 2 (82-138); 946, 1 (48); 947, 1 (131); 950, 1 (61); 953, 1 (49); 998, 6 (127-155); 1006, 4 (88-153); 1007, 2 (110-150); 1010, 3 (69-118); 1015, 1 (73); 1019, 4 (47-135); 1020, 3 (83-138); 1022, 7 (71-140); 1023, 3 (79-117); 1026, 1 (74); 1027, 3 (84-141); 1029, 1 (150); 1030, 1 (112); 1038, 6 (96-158); 1051, 2 (81-135); 1057, 2 (61-64); 1064, 4 (76-127); 1065, 1 (144); 1071, 1 (97); 1076, 2 (131-149); 1077, 5 (80-155); 1121, 2 (29-33); 1132, 1 (93); 1141, 1 (70); 1204, 2 (35-36); 1206, 5 (35-36); 1215, 1 (34); 1224, 1 (34); 1234, 1 (50); 1236, 2

(93-105); 1262, 1 934); 1270, 1 (31); 1324, 2
124-139); 1358, 1 (130); 1359, 2 (85-135); 1384,
2 (46-145); 1393, 4 (46-66); 1454, 4 (32-35);
1463, 2 (128-129); 1507, 1 (128); 1510, 2
(37-41); 1516, 3 (37-142); 1518, 1 (140); 1546, 2
(86-98); 1550, 4 (110-157); 1568, 2 (71-129);
1580, 2 (68-70); 1584, 1 (35); 1645, 1 (46);
1646, 3 (41-45); 1648, 4 (30-47); 1649, 1 (32);
1653, 1 (42); 1658, 1 (44) 1665, 1 (43); 1666, 1
(50); 1685, 1 (63); 1689, 2 (30-33); 1865, 2
(97-129); 1868, 2 (154-164); 1925, 1 (117); 1936,
3 (97-138); 1959, 2 (103-152); 1966, 3 (81-109);
2122, 2 (88-146); 2136, 4 (73-130); 2168, 1
(138); 2174, 1 (53); 2179, 1 (47); 2191, 1 (149);
2234, 1 (34); 2242, 1 (45); 2254, 1 (38); 2293, 2
(38-40); 2294, 1 (38).

    Eltanin (SOSC). 49, 2 (87-107); 54, 1 (85); 63,
1 (80); 69, 1 (52); 72, 1 (100); 80, 4 (30-32);
129, 2 (72-82); 144, 2 (36-152); 145, 1 (133);
151, 1 (46); 156, 1 (150); 157, 1 (37); 352, 3
(35-40); 363, 1 (35).

    Discovery. 151, 1 (129); 1038, 1 (77); 1146, 1
(89); 1298, 1 (77); 1633, 1 (140); 1718, 1 (60);
1869, 1 (122); 2171, 1 (151); 2543, 1 (96); 2594,
1 (?).

    BANZARE. 96, 1 (118); 32, 1 (99).

Gymnoscopelus aphya

    Eltanin (USC). 99, 2 (25-26); 236, 15 (24-45);
246, 3 (25-41); 252, 1 (25); 253, 2 (25-27); 310,
7 (26-29); 313, 38 (22-29): 318, 4 (27-124); 319,
3 (29-30); 348, 20 (26-32); 354, 4 (27-29); 355,
6 (25-30); 359, 33 (27-29); 360, 12 (26-27); 361,
11 (27-29); 375, 95 (26-28); 378, 27 (26-125);
379, 12 (29-31); 381, 4 (28-30); 382, 1 (30);
383, 5 (30-31); 388, 6 (27-30); 392, 2 (28-29);
396, 1 (124); 414, 1 (138); 489, 1 (145); 570, 1
(37); 655, 1 (133); 730, 1 (51); 767, 5 (26-30);
775, 6 (22-28); 778, 1 (25); 846, 5 (28-29); 847,
16 (22-30); 849, 10 (25-30); 850, 2 (28-30);
852, 5 (28-29); 866,1 (29); 867, 2 (30-31); 868,
53 (29-31); 874, 7 (28-30); 877, 7 (28-30); 878,
4 (25-30); 882, 1 (31) 883, 2 (31-32); 885, 12
(31-34); 886, 2 (28-29); 888, 2 (29-30); 889, 1
(30); 1029, 1 (153); 1106, 1 (125); 1142, 1
(136); 1269, 3 (24-26); 1298, 49 (21-31); 1315, 1
(43); 1316, 2 (40-43); 1333, 1 (42); 1336, 1
(150); 1337, 1 (29); 1341, 8 (24-28); 1342, 2
(29-30); 1348, 2 (30-31); 1379, 2 (29-30); 1462,
2 (42-144); 1501, 2 (30-32); 1503, 4 (31); 1518,
1 (34); 1519, 1 (31); 1521, 2 (29-32); 1522, 5
(30-32); 1525, 1 (52); 1526, 1 (39); 1616, 1
(37); 1622, 1 (33); 1648, 1 (36); 1649, 1 (32);
1657, 3 (34-37); 2293, 1 (53); 2299, 1 (124).

    Eltanin (SOSC). 11, 9 (33-39); 14, 1 (55); 41,
3 (139-144); 157, 1 (27); 159, 8 (23-25); 162, 1
(23).

    ELtanin Cruise 21. 21, 2 (30-32).
    Eltanin Cruise 33 (ANARE). 33-5, 1 (145).
    Discovery. 60, 1 (39); 62, 1 (41); 633, 1
(50); 753, 1 (138); 819, 2 (99-109); 1023, 1 (?);
1125, 1 (67); 1233, 1 (34); 1325, 1 (39); 1367, 1
(40); 1492, 1 (135); 1969, 1 (34); 1975, 1 (33).
    William Scoresby. 213, 6 (44-77); 216, 1 (49);
236, 14 (45-73); 236, 13 (43-79).
    British Museum of Natural History. 1873-8-1-42
(type from Godeffroy Museum), 1 (27); 1923-7-19-1
(from stomach of Blue Whale, Deception Island), 1
(145); 1923-7-19-2 (Bransfield Strait), 1 (?).

Gymnoscopelus bolini

    Eltanin (USC). 99, 1 (30). 109, 1 (92); 148, 1
(78); 279, 1 (160); 313, 1 (25); 355, 1 (34);
379, 2 (65-72); 626, 1 (160); 634, 1 (205); 755,
1 (203); 858, 1 (245); 864, 1 (233); 867, 1
(150); 892, 1 (153); 1100, 2 (42-45); 1170, 21
(28-30); 1196, 6 (45-58); 1235, 1 (210); 1269, 2
(63-149); 1270, 1 (30); 1288, 1 (195); 1333, 6
(31-73); 1336, 3 (31-59); 1337, 1 (30); 1342, 1
(31); 1365, 1 (107); 1379, 36, (29-32); 1380, 3
(29-30); 1405, 1 (25); 1443, 1 (77); 1463, 1
(147); 1470, 1 (58); 1475, 1 (90); 1481, 1 (47);
1522, 2 (112-188); 1586, 2 (107-199); 1603, 1
(109); 1607, 1 (140); 1734, 1 (38); 1834, 1 (26);
2218, 1 (53); 2233, 1 (58); 2245, 1 (60); 2247, 1
(53); 2268, 3 (30-174).

    Eltanin (SOSC). 11, 3 (42-53); 92, 7 (23-29);
340, 11 (24-29); 344, 1 (29); 348, 5 (31-32);
349, 1 (30).

    Eltanin Cruise 21. 21-21, 1 (34); 21-16, 1
(36).

    Discovery. 106, 1 (72); 217, 1 (69); 751, 1
(96); 1916, 1 (87).

Gymnoscopelus fraseri

    Eltanin (USC). 99, 13 (30-42); 109, 2 (53-68);
149, 1 (52); 246, 1 (63); 306, 1 (63); 318, 2
(68-78); 354, 1 (34); 563, 2 (40-42); 567, 1
(70); 571, 1 (69); 588, 1 (67); 593, 2 (76-80);
667, 5 (64-79); 670, 1 (60); 730, 1 (66); 738, 1
(76); 781, 2 (65-78); 846, 1 (69); 847, 1 (37);
850, 1 (35); 868, 1 (34); 877, 6 (49-69); 878, 3
(39-44); 882, 4 (34-55); 883, 3 (46-51); 888, 6
(33-48); 890, 2 (68-80); 892, 2 (55-70); 918, 1
(69); 957, 1 (57); 1100, 7 (53-66); 1107, 1 (35);
1114, 1 (62); 1119, 1 (73); 1142, 2 (46-52);
1167, 1 (72); 1196, 18 (31-53); 1204, 6 (31-77);
1206, 1 (67); 1210, 2 (29-45); 1220, 1 (70);
1241,. 10 (43-68); 1262, 2 (42-45); 1269, 1 (48);
1270, 3 (34-38); 1294, 2 (41-62); 1295, 4
(64-69); 1299, 5 (28-76); 1303, 3 (31-60); 1304,
4 (33-64); 1306, 1 (62); 1307, 4 (31-57)' 1315, 2
(29-71); 1316, 5 (33-78); 1319, 1 (60); 1323, 1
(49); 1324, 3 (42-85); 1333, 6 (37-50); 1336, 1
(31); 1337, 9 (32-68); 1342, 4 (37-78); 1348, 17
(46-79); 1350, 1 (85); 1364, 1 (50); 1365, 1
(66); 1374, 1 (83); 1379, 17 (26-67); 1380, 10
(24-33); 1392, 1 (66); 1396, 1 (27); 1409, 1
(35); 1439, 22 (36-75); 1454, 4 (60-80); 1456, 2
(57-65); 1462, 1 (70); 1463, 1 (67); 1471, 5
(56-76); 1473, 5 (38-75); 1475, 1 (53); 1481, 2
(30-32); 1501, 1 (65); 1516, 1 (66); 1525, 1
(62); 1528, 1 (74); 1590, 2 (64-68); 1606, 4
(55-61); 1607, 5 (50-59); 1608, 2 (47-53); 1622,
1 (79); 1636, 1 (56); 1637, 2 (57-75); 1645, 17
(56-79); 1648, 1 (55); 1653, 1 (36); 1661, 1
(58); 1662, 1 (54); 1666, 2 (62-67); 1677, 2
(60-64); 1684, 1 (84); 1685, 2 (43-70); 1686, 5
(29-81); 1687, 2 (61-67); 1689, 1 (33); 1834, 1
(28); 1835, 1 (30); 1840, 12 (30-34); 1977, 4
(34-47); 2179, 1 (70); 2183, 2 (62-65); 2187, 3
(59-63); 2191, 2 (29-38); 2204, 1 (45); 2205, 2
(30-32); 2208, 1 (42); 2211, 1 (45); 2212, 1
(65); 2213, 4 (50-59); 2214, 2 (30-41); 2217, 7
(26-36); 2218, 12 (28-41); 2236, 1 (74); 2242, 1
(61); 2245, 2 (54-57); 2246, 2 (32-61); 2257, 1
(32); 2260, 1 (42); 2265, 4 (45- 76); 2266, 1
(45); 2283, 2 (25-26); 2289, 1 (80); 2290, 1

(37); 2291, 1 (61); 2292, 1 (83); 2295, 1 (77); 2298, 1 (65).

Eltanin (SOSC). 11, 15 (30-61); 16, 1 (61); 28, 1 (67); 41, 1 (71); 69, 1 (78); 71, 3 (53-70); 74, 4 (54-58); 76, 4 (49-71); 159, 1 (?33); 168, 2 (?36-41); 328, 2 (35-36); 330, 1 (30); 332, 1 (33); 338, 5 (35-40); 340, 1 (33); 341, 1 (35); 344, 3 (29-43); 345, 1 (38); 349, 6 (36-78); 350, 1 (75); 351, 1 (33); 352, 7 (32-45); 354, 14 (34-70); 362, 4 (52-84).

Discovery. 357, 1 (72); 665, 2 (62-64); 670, 1 (42); 750, 1 (75); 775, 1 (55); 838, 1 (28); 867, 1 (40); 956, 1 (69); 1189, 1 (27); 1320, 1 (59); 1339, 1 (80); 1440, 1 (72); 1614, 1 (46); 1682, 1 (36); 1677, 1 (70); 1747, 1 (44); 2067, 1 (25); 2218, 1 (59); 2289, 1 (66); 2496, 1 (64).

William Scoresby. 575, 1 (79)

British Museum of Natural History. 1931-2-27-6 (holotype from 3°18'S, 5°17'E), 1 (50).

Gymnoscopelus sp. A

Eltanin (USC). 97,3 (42-49); 99, 2 (24-28); 169, 2 (63-66); 882, 1 (65); 957, 1 (28); 1100, 9 (57-63); 1142, 1 (52); 1167, 1 (50); 1241, 1 (69); 1285, 1 (48); 1355, 1 (51); 1439, 21 (28-60); 1684, 1 (57); 1728, 1 (74); 1777, 1 (68); 1798, 2 (69-70).

Eltanin (SOSC). 11, 3 (31-62); 85, 1 (61); 315, 1 (24); 316, 1 (67); 317, 126 (21-54); 321, 1 (50); 327, 5 (42-48); 328, 12 (24-41); 336, 1 (45); 340, 4 (26-27); 341, 2 (45-50); 344, 2 (25-60); 345, 2 (48-53); 348, 1 (27); 349, 5 (52-67).

Eltanin Cruise 21. 21-16, 5 (25-67).

Discovery. 969, 1 (76); 2738, 1 (53).

Scripps Institution of Oceanography. SIO 61-45-25 (46°53'S, 179°48.32'W), 1 (65).

Gymnoscopelus sp. B

Eltanin (USC). 97, 8 (42-59); 165, 4 (40-111); 169, 4 (50-52); 215, 7 (40-50); 252, 1 (86); 306, 1 (96); 319, 1 (78); 778, 1 (70); 847, 1 (75); 849, 1 (38); 852, 1 (97); 868, 2 (71-104); 882, 1 (47); 883, 1 (85); 914, 1 (98); 950, 1 (89); 1107, 1 (60); 1132, 1 (99); 1162, 1 (117); 1167, 2 (75-125); 1187, 1 (73); 1270, 1 (80); 1307, 1

(68); 1380, 2 (49-52); 1645, 3 (86-89); 1648, 1 (93); 1653, 2 (84-85); 1661, 3 (67-85); 1671, 1 (95); 1685, 2 (71-72); 1695, 2 (39-46); 1696, 1 (43); 1823, 1 (32); 1831, 1 (29); 1838, 1 (31); 1959, 1 (109); 2168, 1 (101); 2234, 1 (134); 2241, 1 (131); 2244, 1 (59); 2247, 1 (50); 2260, 1 (71); 2266, 1 (54); 2269, 1 (42).

Eltanin (SOSC). 63, 1 (107); 82, 1 (29); 348, 2 (48-53); 350, 1 (52).

Eltanin Cruise 21. 21-16, 5 (54-60); 21-21, 1 (114).

Discovery. 256, 2 (32-33); 837, 1 (30); 976, 1 (71); 1299, 1 (102); 2737, 1 (45).

William Scoresby. 587, 2 (42-50).

Gymnoscopelus piabilis

Eltanin (USC). 97, 16 (42-53); 99, 1 (42); 165, 50 (22-91); 169, 8 (24-54); 175, 1 (21); 213, 3 (22-23); 215, 9 (26-54); 336, 1 (72); 563, 3 (22-23); 667, 1 (38); 741, 1 (105); 957, 2 (27); 1607, 1 (32); 2203, 1 (34); 2217, 2 (35-46); 2218, 3 (41-52); 2219, 1 (43); 2221, 2 (43-87); 2226, 1 (90).

Eltanin (SOSC). 2288, 1 (91).

Discovery. 1189, 1 (69).

William Scoresby. 587, 5 (38-45).

Gymnoscopelus sp. C

Eltanin (USC). 378, 1 (93); 775, 1 (105); 874, 1 (102); 1106, 1 (96); 1113, 2 (80-83); 1162, 1 (108); 1167, 1 (77); 1186, 3 (97-104); 1405, 2 (63-65); 1409, 14 (24-29); 1432, 48 (22-103); 1692, 1 (30); 1695, 3 (29-30); 1700, 2 (29-32); 1701, 1 (28); 1728, 1 (31); 1838, 5 (58-90); 1841, 2 (91-102); 1977, 1 (110); 2191, 1 (69); 2216, 1 (30); 2218, 1 (32); 2226, 19 (23-33); 2245, 1 (86); 2247, 1 (70); 2252, 2 (33-35); 2268, 2 (33-35); 2269, 7 (26-35); 2270, 5 (27-37); 2283, 1 (33); 2285, 3 (34-37); 2286, 1 (69); 2287, 2 (70-72); 2288, 4 (70-104); 2300, 2 (70-90); 2302, 2 (30-39).

Eltanin (SOSC). 349, 2 (37-39).

Anton Bruun Cruise 3 [Nafpaktitis and Nafpaktitis, 1969]. 7127 (40°53'S, 60°01'E), 1 (109).

Appendix 2. Species Involved in Patterns of Distribution

Species involved in major patterns and subpatterns of distribution proposed for south of 30°S are given here. Species marked with asterisks are based on literature records only.

Pattern 1

Subpattern A. Protomyctophum (Hierops) chilensis. P. (Hierops) sp. D, Symbolophorus sp. D, Diaphus danae.

Subpattern B. Diaphus parri, Myctophum phengodes, Lampadena dea, Scopelopsis multipunctatus, Gonichthys barnesi, Hygophum hygomi, Lampadena speculigera, L. chavesi,* Lampanyctus pusillus, Diaphus effulgens.

Subpattern C. Lepidophanes guentheri, Lampanyctus parvicauda,* Triphoturus mexicanus*.

Subpattern D. Electrona rissoi, Benthosema suborbitale, Diogenichthys atlanticus, Notolychnus valdiviae, Ceratoscopelus warmingi, Taaningichthys bathyphilus, Lampanyctus alatus, Lobianchia dofleini, Notoscopelus resplendens.

Subpattern E. Species incertae cedis. Hygophum macrochir group, Myctophum nitidulum,* Symbolophorus spp., Centrobranchus nigroocellatus,* Bolinichthys supralateralis, B. indicus,* Lampanyctus niger-ater group, L. sp. B, L. sp. C, Lobianchia gemellari,* Diaphus termophilus, D. mollis, D. sp. B.

Pattern 2

Subpattern A. Protomyctophum (Hierops) subparallelum, Metelectrona sp. A, Hygophum hanseni, Lampanyctus sp. A, L. intricarius, L. sp. D, Diaphus ostenfeldi, D. sp. A, Hintonia candens, Loweina interrupta.

Subpattern B. Electrona paucirastra, Metelectrona ventralis, Symbolophorus sp. B, Lampadena notialis, Lampanyctus australis, L. lepidolychnus, Lampichthys procerus, Hygophum brunni, Symbolophorus sp. C, Lampanyctus iselinoides, Lampanyctodes hectoris.

Pattern 3

Protomyctophym normani, P. sp. C, P. (Hierops) parallelum, Electrona subaspera, Symbolophorus sp. A, Lampanyctus achirus, L. macdonaldi, Gymnoscopelus sp. A, G. piabilis, G. sp. C.

Pattern 4

Protomyctophum tenisoni, P. sp. A, P. andriashevi, P. sp. B, Electrona carlsbergi, Gymnoscopelus bolini, G. sp. C.

Pattern 5

Protomyctophum anderssoni, P. bolini, Electrona antarctica, Gymnoscopelus braueri, G. opisthopterus, G. aphya, G. sp. B.

References

Addicott, W. O.
  1970 Latitudinal gradients in tertiary mol-
        luscan faunas of the Pacific coast.
        Palaeogeogr. Palaeoclimatol. Palaeoe-
        col., 8: 287-312.
Ahlstrom, E. H.
  1971 Kinds and abundance of fish larvae in
        the eastern tropical Pacific, based on
        collections made on Eastropac I. Fish.
        Bull., 69: 3-77.
Alcock, A. W.
  1891 Natural history notes from H.M. Indian
        Marine Survey Steamer Investigator. No.
        16. On the bathybial fishes collected
        in the Bay of Bengal during the season
        1889-90. Ann. Mag. Nat. Hist., ser. 6,
        6: 197-222.
Andrews, K. J. H.
  1966 The distribution and life history of
        Calanoides acutus. Discovery Rep., 34:
        177-162.
Andriashev, A. P.
  1958 List of ichthyological stations with
        preliminary characterization of the hauls
        (in Russian). Trudy Kompl. Antarkt.
        Eksped. Akad. Nauk SSSR, Gidrol. Gidrokh-
        him. Geol. Biol. Issled. D/E/ Ob 1955-
        1956: 199-204.
  1961 List of ichthyological stations with
        preliminary characterization of the hauls
        (in Russian). Trudy Sov. Antarkt. Ek-
        sped., 22: 227-234.
  1962 Bathypelagic fished of the Antarctic. 1.
        Family Myctophidae (in Russian). Izv.
        Akad. Nauk SSSR Fauny Morei, 1(9): 216-
        294.
Angelescu, V., and M. B. Cosseau.
  1969 Alimentacion de las merluza en la region
        del Talud Continental Argentino, epoca
        internal (Merluciidae, Merlucius mer-
        lucius hubbsi). Boln. Inst. Biol. Mar.,
        19: 1-91.
Backus, R. H., G. W. Mead, R. L. Haedrich,
and A. W. Ebling
  1965 The mesopelagic fishes collected during
        cruise 17 of the R/V Chain with a method
        of analyzing faunal transects. Bull.
        Mus. Comp. Zool. Harv., 145: 139-158.
Backus, R. H., J. E. Craddock, R. L. Haedrich,
and D. L. Shores
  1969 Mesopelagic fishes and thermal fronts in
        the western Sargasso Sea. Mar. Biol., 3:
        87-106.
  1970 The distribution of mesopelagic fishes
        in the equatorial and western North At-
        lantic Ocean. J. Mar. Res., 28: 179-201.
Baker, A. de C.
  1954 The circumpolar continuity of Antarctic
        plankton species. Discovery Rep., 27:
        201-218.
  1959 Distribution and life history of Euph-
        ausiia triacantha Holt and Taterstal.

Discovery Rep., 29: 309-340.
  1965 The latitudinal distribution of Euph-
        ausia species in the surface waters of
        the Indian Ocean. Discovery Rep., 33:
        309-334.
Balech, E.
  1970 The distribution and endemism of some
        Antarctic microplankton. In M. W. Hold-
        gate (Ed.), Antarctic ecology, I: 143-
        146. Academic, London.
Barnard, K. H.
  1925 A monograph of marine fishes of South
        Africa. 1. Ann. S. Afr. Mus., 21: 235-
        247.
Barsukov, V. V., and Y. E. Permitin
  1959 List of ichthyological collections (in
        Russian). Trudy Sov. Antarkt. Eksped.
        6: 379-387.
Be', A. W. H.
  1969 Planktonic Foraminifera. Antarct. Map
        Folio Ser., folio 11: 9-12, pls 1-2.
        Amer. Geogr. Soc., New York.
Becker, V. E.
  1963a New data on the lanternfish genera Elec-
         trona and Protomyctophum (Pisces, Mycto-
         phidae) of the southern hemisphere (in
         Russian). Vop. Ikhtiol., 3: 15-28.
  1963b North Pacific species of the genus Proto-
         myctophum (Myctophidae, Pisces) (in Rus-
         sian). Trudy Inst. Okeanol., 62: 164-
         191.
  1964a Slendertailed lanternfishes (Genera Low-
         eina, Tarletonbeania, Gonichthys, and
         Centrobranchus) of the Pacific and In-
         dian Oceans. Systematics and distrib-
         ution (in Russian). Trudy Inst. Oke-
         anol., 73: 11-75.
  1964b The temperate-cold water myctophid com-
         plex (Myctophidae, Pisces) (in Russian).
         Okeanologiya, 4(3): 464-476.
  1965   Lanternfishes of the genus Hygophum.
         Systematics and distribution (in Rus-
         sian). Trudy Inst. Okeanol., 80:62-103.
  1967a Lanternfishes (Family Myctophidae) from
         the Petr Lebedev Atlantic expeditions
         1961-1964 (in Russian). Trudy Inst.
         Okeanol., 84: 84-124.
  1967b Lanternfishes (Family Myctophidae) (in
         Russian). In T. S. Rass (Ed.), Biology
         of the Pacific Ocean, Book 3: 145-181.
         Nauka, Moscow.
Beebe, W.
  1937 Preliminary list of Bermuda deepsea fish-
        es. Zoologica, 22: 197-208.
Beebe, W., and M. Van der Pyl
  1943 Eastern Pacific Expedition of the New
        York Zoological Society XXXIII. Pacific
        Myctophidae (Fishes). Zoologica, 29:
        59-95.
Beklemishev, K. V.
  1967 Biogeographical division of the pelagial
        of the Pacific Ocean (in surface and

intermediate waters) (in Russian). Tik-
hiy Okean, VII. Biology of the Pacific
Ocean, 1. Nauka, Moscow.

Berry, F. H. and H. C. Perkins
1966   Survey of pelagic fishes of the California Current area. Fish. Bull., 65;
625-682.

Blache, J., and A. Stauch
1964a Contribution a la connaissance des poissons de la famille Myctophidae dans la
partie orientale du Golfe de Guiness
(Teleostei, Clupeiformi, Myctophidae).
1. Les genres Electrona Goode and Bean
1895, Hygophum Bolin 1939. Cah. Oceanogr., 1: 97-104.
1964b Contribution a la connaissance des poissons de la famille Myctophidae dans la
partie orientale du Golfe de Guiness
(Teleostei, Clupeiformi, Myctophidae).
2. Les genres Diogenichthys.. Cah.
Oceanogr., 2(2): 61-78.

Blackburn, M.
1968   Micronekton of the eastern tropical Pacific Ocean: Family composition, distribution, avoidance, and relation to tuna.
Fish. Bull., 67(1): 71-115.

Boisvert, W. E.
1967   Major currents in the North and South
Atlantic Oceans between 64°N and 60°S.
Tech. Rep. 193: 1-92. Nav. Oceanogr.
Office, Washington, D. C.

Bolin, R. L.
1939   A review of the myctophid fishes of the
Pacific coast of the United States and
of lower California. Stanford Ichthyol.
Bull., 1: 89-156.
1946   Lanternfishes from Investigator station
670, Indian Ocean. Stanford Ichthyol.
Bull., 1: 89-156.
1959   Iniomi. Myctophidae. Report of the Scientific results of Michael Sars, North
Atlantic Deep-Sea Expedition, 1910, 4
(7): 1-45. Bergen Museum, Bergen, Nor-
way.

Boltovskoy, E.
1968   Hidrologia de las aguas suferficiales en
la parte occidental del Atlantic sur.
Hydrobiologia, 2(6): 199-224.

Bonaparte, C. L. J. L.
1840   Iconograffia della fauna italica per la
quatre classi degli anamali vertebrati.
vol. 3, Pesci. Roma.

Botnikov, V. N.
1966   The limits of the West Wind Drift. Inf.
Bull. Sov. Antarkt. Eksped., 6(1): 48-
51.

Boulenger, G. A.
1902   Pisces. Report on the collections of
natural history made in Antarctic regions
during the voyage of the Southern Cross,
2: 174-189.

Brauer, A.
1904   Die Gattung Myctophum. Zool. Anzeiger,
28(10): 377-404.
1906   Die Tiefsee Fische. 1. Systematischer
Teil. Wiss. Ergeben. Dt. Tiefsee Exped.
Valdivia 1898-1899, 15(1): 1-432.

Briggs, J.
1966   Oceanic Islands, endemism, and marine
paleotemperatures. System. Zool. 15(2):
153-163.

1970   A faunal history of the North Atlantic
Ocean. Syst. Zool., 19(1): 19-34.

Brinton, E.
1962   The distribution of Pacific Euphausiids.
Bull. Scripps Instn. Oceanogr., 8(2): 51-
270.

Brodskey, K. A.
1965   The taxonomy of marine plankton organisms and oceanography. Oceanology, 5(4):
1-11.

Burling, R. W.
1961   Hydrology of circumpolar waters south of
New Zealand. N.Z. Dep. Sci. Ind. Res.
Bull., 143: 1-66.

Bussing, W. A.
1965   Studies of the midwater fishes of the
Peru-Chile Trench. In G. A. Llano (Ed.),
Biology of the Antarctic Seas 2. Ant-
arctic Res. Ser., 5: 185-227. AGU,
Washington, D. C.

Chapman, W. M.
1944   A new name for Myctophum oculeum Chapman. Copeia, 1: 54-55.

Cocco, A.
1829   Su di alcuni pesci Mari di Messina.
Giorn. Sci. Lett. Arti Sicilia, 26:
138-147. Palermo.
1838   Su di alcuni Salmonidi del Mar di Messina, lettera al C. D. C. L. Bonaparte.
Nuovi. Ann. Sci. Nat. Bolgna, 2: 161-
194.

Cohen, D. M.
1970   How many recent fishes are there? In
Festschrift for George Sprague Myers in
honor of his sixty-fifth birthday. Proc.
Calif. Acad. Sci., 38(4): 341-345.

Coleman, L. R., and B. G. Nafpaktitis
1972   Dorsadena yacquinae, a new genus and
species of myctophid fish from the eastern North Pacific Ocean. Nat. His. Mus.
Los Angeles, County Contr. Sci., 225:
1-11.

Craddock, J. E., and G. W. Mead
1970   Midwater Fishes from the eastern south
Pacific Ocean. Sci. Res. SE Pacif. Exped. Anton Bruun Rep., 3: 3-46.

Danilchenko, P. G.
1967   Bony fishes of the Maikop deposits of
the Caucasus. Translated from Russian,
catalogue 1885. Israel Program for Scientific Translations, Jerusalem.

David, P. M.
1955   The distribution of Sagitta gazelle Rit-
ter-Zahoney. Discovery Rep., 27: 235-
278.
1958   The distribution of the Chaetognatha of
the Southern Ocean. Discovery Rep.,
29: 201-208.
1963   Some aspects of speciation on Chaetognatha. Publ. 5: 129-143. Speciation
in the Sea. Syst. Ass., London.

Davy, B.
1972   A review of the lanternfish genus Taaningichthys (Family Myctophidae) with the
description of a new species. Fish.
Bull., 70(1): 67-68.

Deacon, G. E. R.
1933   A general account of the hydrology of
the South Atlantic Ocean. Discovery
Rep., 7: 171-238.
1937   The hydrology of the Southern Ocean.

Discovery Rep., 15: 1-124.

1963   The Southern Ocean, In M. N. Hill (Ed.), Ideas and observations on progress in the study of the seas, The Sea, 2: 28-296. John Wiley, New York.

Defant, A.
1961   Physical oeanography. Vol. 1 729 pp. Pergamon, New York.

Devereaux, I.
1967   Oxygen isotope paleotemperature measurements on New Zealand Tertiary fossils. N.Z. J. Sci., 10(4): 984-1011.

Dewitt, H. H.
1970   The character of the midwater fish fauna of the Ross Sea, Antarctica, In M. W. Holdgate (Ed)., Antarctic ecology, 1: 305-314, Academic, London.

Dewitt, H. H., and J. C. Taylor
1960   Fishes of the Stanford Antarctic biological research program 1958-1959. Stanford Ichthyol. Bull., 7(4): 162-199.

Dietrich, C.
1963   General oceanography. 588 pp. John Wiley, New York.

Ebeling, A. W.
1962   Melamphaidae. I. Systematics and zoogeography of the species in the bathypelagic fish genus Melamphaes Günther. Dana Rep., 58: 1-159.

1967   Zoogeography of tropical deep sea animals. Proc. Int. Conf. Tropical Oceanogr., 5: 593-613.

Eigenmann, C. H. and R. S. Eigenmann
1889   Notes from the San Diego Biological Laboratory. The fishes of Cortez Banks. West. Amer. Sci., 6(48): 123-132.

1890   Addition to the fauna of San Diego. Proc. Calif. Acad. Sci., Ser. 2, 3: 1-24.

Ekman, S.
1953   Zoogeography of the sea. 417 pp. Sidgewick and Jackson, London.

El-Sayed, S. Z.
1970   Phytoplankton production of the South Pacific and the Pacific sector of the Antarctic, Scientific Exploration of the South Pacific, W. S. Wooster (Ed.), 194-210. National Academy of Sciences, Washington, D. C.

Esteve, R.
1947   Revision des types myctophides (Scopelides) du Museum. Bull. Mus. Hist. Nat. Paris., Ser. 2, 19(1): 67-69.

Fell, H. B.
1967   Cretaceous and tertiary surface currents of the oceans, In H. Barnes (Ed.), Oceanography and marine biology, Annual Review 5: 317-341, George Allen and Unwin, London.

Fitch, J. E.
1969   Fossil lanternfish otoliths of California, with notes on fossil myctophidae of North America. Los Angeles County Mus., Contrib. Sci., 173: 1-20.

Fitch, J. E., and R. Brownell
1968   Fish otoliths in cetacean stomachs and their importance in interpreting feeding habits. J. Fish. Res. Bd. Can., 25: 2561-2574.

Fleming, C. A.
1962   New Zealand biogeography, a paleontologist's approach. Tuatara, 10: 53-108.

Fleminger, A.
1967   Taxonomy, distribution, and polymorphism in the Labidocera jollae group with remarks on evolution within the group (Copepoda: Calanoidea). Proc. U.S. Natnl. Mus., 120(3567): 1-61.

Fowler, H. W.
1901   Myctophum phengodes in the North Atlantic. Proc. Acad. Nat. Sci. Philad., 53: 620-621.

1925   New taxonomic names of West African marine fishes. Am. Mus. Novit., 162: 1-5.

Foxton, P.
1956   The distribution of the standing crop of zooplankton in the Southern Ocean. Discovery Rep., 28: 191-236.

1961   Salpa fusiformis and related species. Discovery Rep., 32: 1-32.

1965   The distribution and life history of Salpa thompsoni Foxton with observations on a related species, Salpa gerlachei Foxton. Discovery Rep., 34: 1-116.

Frakes, L. A., and E. M. Kemp
1972   Influence of continental positions on early Tertiary climates. Nature, 240: 97-100.

Fraser-Brunner, A.
1931   Some interesting West African fishes, with descriptions of a new genus and two new species. Ann. Mag. Nat. Hist., Ser. 10, 8: 217-225.

1949   A classification of the fishes of the Family Myctophidae. Proc. Zool. Soc. London, 118(4): 1010-1106.

Friedman, S. B.
1964   Physical oceanographic data obtained during Eltanin cruises 4, 5 and 6 in the Drake Passage, along the Chilean coast and in the Bransfield strait, June, 1962-Janury, 1963. Tech. Rep. 1 Cu-1-64: 55 pp. Lamont-Doherty Geol. Observ., Palisades, N. Y.

Frost, B., and A. Fleminger
1968   A revision of the Genus Clausocalanus (Copepoda: Calanoidea) with remarks on distribution patterns in diagnostic characters. Bull. Scripps Instn. Oceanogr., 12: 1-235.

Garman, S.
1899   Reports of an exploration off the west coasts of Mexico, Central and South America, and off the Galapagos Islands, in charge of Alexander Agassiz, by the U.S. Fish Commission Steamer Albatross, during 1891, Lieut. Commander Z. L. Tanner U. S. N., commanding, XXVI. The Fishes. Mem. Mus. Comp. Zool. Harv., 24: 1-431.

Garner, D. M.
1967a   Hydrology of the southwest Tasman Sea. Mem. N.Z. Oceanogr. Inst., 48: 1-40.

1967b   Hydrology of the southern Hikurangi trench region. Mem. N.Z. Oceanogr. Inst., 34: 1-33.

Gatti, M. A.
1903   Richerche sugli ovgawi luminosi dei pesci. Annali Agric. Roma, 233: 1-126.

Gibbs, R. H.
1968   Photonectes munificus, a new species of melanostomatid fish from the south Pac-

ific Subtropical Convergence, with remarks on the convergence Fauna. Los
Angeles County Mus., Contrib. Sci., 149:
1-6.

Gilbert, C. H.
  1891 Preliminary report on the fishes col-
         lected by the steamer Albatross on the
         Pacific coast of North America during
         the year 1889. Proc. U.S. Natn. Mus.,
         13: 49-126.
  1905 The deep sea fishes of the Hawaiian Is-
         lands. Bull. U.S. Fish. Comm., 2:
         575-713.
  1911 Notes on lanternfishes from Southern
         Seas, collected by J. T. Nichols in 1906.
         Bull. Am. Mus. Nat. Hist., 30: 13-19.
  1913 The lanternfishes of Japan. Mem. Car-
         neg. Mus., 6: 67-107.

Gilbert, C. H. and F. Kramer
  1897 Report on the fishes dredged in deep
         water near the Hawaiian Islands, with
         descriptions and figures of twenty three
         new species. Proc. U.S. Natl. Mus.,
         19: 403-435.

Gilchrist, J. D. F.
  1904 South African fishes. Mar. Invest. S.
         Afr., 3: 15.

Gistel, J.
  1850 Gonichthys, ein fisch aus der Bai von
         Madeira Isis, 1850: 71-72.

Goode, G. B., and T. H. Bean
  1896 Oceanic ichthyology. Spec. Bull. U.S.
         Natn. Mus., 2: 1-553.

Goody, P. C.
  1969 The relationships of certain Upper Cre-
         taceous telecosts with special reference
         to the Myctophoids. Bull. Br. Mus. Nat.
         Hist. Geol., 7: 1-255.

Gordon, A. L.
  1967 Structure of antarctic waters between
         20°W and 170°W. Amer. Geogr. Soc., New
         York. Antarctic Map Folio Ser. folio
         6: 1-10, pls. 1-14.
  1971a Oceanography of antarctic waters. In J.
         L. Reid (Ed.), Antartic Oceanology 1,
         Antarctic Res. Ser., 15: 169-208.
  1971b Antarctic polar front zone. In J. L.
         Reid (Ed.), Antartic Oceanology, Ant-
         arctic Res. Ser., 15: 205-221. AGU,
         Washington, D. C.

Gordon, A. L., and R. D. Goldberg
  1970 Circumpolar characteristics of Antartic
         waters. Amer. Geogr. Soc., New York.
         Antarctic Map Folio Ser. folio 13: 1-5,
         pls. 1-18.

Grandperrin, R., and J. Rivaton
  1966 Coriolis Croisiere Alize. Individual-
         ization de plusiers ichtyofaunes le long
         de l'equater. Cah. Oceanogr., 4(4):
         35-49.

Guenther, F. R.
  1936 A report on oceanographical investiga-
         tions in Peru Coastal Current. Dis-
         covery Rep., 13: 107-276.

Günther, A. L. C. G.
  1864 Catalogue of the fishes in the British
         Museum. Vol. 5. 455 pp. British Mu-
         seum, London.
  1873 Zweiter ichthyologischer Beitrag nach
         Exemplaren aus dem Museum Godeffroy. J.
         Mus. Godeffroy, 1(4): 256-268.
  1876 Remarks on fishes with descriptions of

new species in the British Museum, chief-
ly from southern seas. Ann. Mag. Nat.
Hist., Ser. 4, 17(43): 389-402.
  1878 Preliminary notices of deep-sea fishes
         collected during the voyage of the H.M.S.
         Challenger, Ann. Mag. Nat. Hist., Ser.
         5, 2(8): 179-187.
  1887 Report on the deep-sea fishes collected
         by H.M.S. Challenger during the years
         1873-1876. Challenger Rep., 22(57):
         1-268.

Hamon, B. V.
  1965 The East Australian Current, 1960-1964.
         Deep Sea Res., 12: 899-921.
  1970 Western boundary currents, Scientific
         Exploration of the South Pacific, W. S.
         Wooster (Ed.), 257 pp. National Academy
         of Sciences, Washington, D. C.

Hardy, A. C.
  1956 The open sea: Its natural history. 1.
         The world of plankton. 350 pp. Collier,
         London.

Hardy, A. C., and E. R. Guenther
  1935 Plankton of the South Georgia whaling
         grounds and adjacent waters. Discovery
         Rep., 11: n 1-456.

Hart, T. J.
  1942 Phytoplankton periodicity in Antarctic
         surface waters. Discovery Rep., 21:
         261-356.

Hasle, G. R.
  1969 An analysis of the phytoplankton of the
         Pacific Southern Ocean: Abundance, com-
         position, and distribution during the
         Brategg Expedition, 1947-1948. Hvalrad.
         Skr., 52: 1-168.

Hays, J. D.
  1968 Climatic record of late Cenozoic Ant-
         arctic Ocean sediments related to the
         record of world climate, Vol. 4. Pale-
         ocology of Africa and of the Surrounding
         Islands and of Antarctica, E. M. van Z.
         Bakker and J. A. Coetzee (Eds.), 139-162.
         Ballama, Capetown.

Hays, J. D., T. Saito, N. D. Opdyke,
and L. H. Burckle
  1969 Pliocene-Pleistocene sediments of the
         equatorial Pacific: Their paleomagnetic,
         biostratigraphic, and climatic record.
         Geol. Soc. Am. Bull., 80: 1481-1514.

Heath, R. A.
  1968 Geostrophic currents derived from oceanic
         density measurements north and south of
         the Subtropical Convergence east of New
         Zealand. N.Z. J. Mar. Freshwater Res.,
         2(4): 659-677.

Herdman, H. F. P.
  1966 Southern Ocean. Encyclopaedia of Ocean-
         ography, R. W. Fairbridge (Ed.), 1021
         pp. Reinhold, New York.

Hubbs, C. L., and R. W. Wisner
  1964 Parvilux, a new genus of myctophid fishes
         from the northeastern Pacific, with two
         new species. Zool. Meded., 39: 445-463.

Hurley, D. E.
  1969 Amphipoda Hyperiidea. Antarctic Map
         Folio Ser. folio 11: 32-34, pls. 18-19.
         Am. Geogr. Soc., New York.

Isaacs, J. D., and L. W. Kidd
  1952 Isaacs-Kidd midwater trawl. Equip. Rep.
         1: 1-18. Scripps Inst. of Oceanogr.,
         La Jolla, Calif.

Jacobs, S. S.
1966   Physical and chemical oceanographic ob-
        servations in the Southern Oceans, U.S.-
        N.S. Eltanin cruises 16-21. Tech. Rep.
        1-CU-66: 1-128. Lamont-Doherty Geol.
        Observ., Palisades, N. Y.
Jacobs, S. S., and A. Amos
1967   Physical and chemical oceanographic ob-
        servations in the Southern Oceans, U.S.-
        N.S. Eltanin cruises 22-27. Tech. Rep.
        1-CU-67: 1-287. Lamont-Doherty Geol.
        Observ., Palisades, N. Y.
John, D. D.
1936   The southern species of the genus Euph-
        ausia. Discovery Rep., 14: 165-180.
Johnson, J. Y.
1890   On some new species of fish from Madeira.
        Proc. Zool. Soc. London, 58: 452-459.
Johnson, M. W., and E. Brinton
1963   Biological species, water-masses and
        currents, In M. N. Hill (Ed.), Ideas and
        observations on progress in the study of
        the seas, The Sea, 2: 381-414. John
        Wiley, New York.
Kane, J. E.
1966   The distribution of Parathemisto gaudi-
        chauddi (Guer.) with observations on its
        life history in the 0° to 20°E sector of
        the Southern Ocean. Discovery Rep.,
        34: 163-198.
Kashkin, N. I.
1967   On the quantitative distribution of the
        lanternfishes (Myctophidae sensulato) in
        the Atlantic Ocean (in Russian). Trudy
        Akad. Nauk SSSR, Inst. Okeanol, 84:
        125-155.
Kemp, S., and A. C. Hardy
1929   The Discovery investigations: Objects,
        equipment, and methods, 2. Discovery
        Rep., 1: 151-222.
Kennett, J. P.
1970   Pleistocene paleoclimates and foramin-
        iferal biostratigraphy in subantarctic
        deep sea cores. Deep Sea Res., 17(1):
        125-140.
Kent, D., N. D. Opdyke, and M. Ewing
1971   Climate change in the North Pacific us-
        ing ice-rafted detritus as climate in-
        dicator. Geol. Soc. Am. Bull., 5(82):
        2741-2754.
Knox, G. A.
1963   Problems of speciation in intertidal
        animals with special reference to New
        Zealand shores. Publ. 5: 129-143.
        Syst. Ass., London.
1970   Biological oceanography of the South
        Pacific, Scientific Exploration of the
        South Pacific, W. S. Wooster (Ed.), 155-
        182. National Academy of Sciences, Wash-
        ington, D. C.
Koblentz-Mishke, O. J., V. S. Volkolnsky,
and J. G. Kaboanova
1970   Plankton primary production of the world
        ocean, Scientific Exploration of the
        South Pacific, W. S. Wooster (Ed.), 183-
        193. National Academy of Sciences, Wash-
        ington, D. C.
Kort, V. G.
1962   The Antarctic Ocean. Scient. Am., 207-
        (3); 113-128.
Kort, V. G., E. S. Kortankavich, and V. G. Bedena
1965   Boundaries of the Southern Ocean (in

Russian). Inf. Byull. Sb. Antarkt. Eksped., 4: 261-263.
Krefft, G.
1958   Antarktische Fische und fischlarven aus
        den Planktonfangen. Dt. Antarkt. Exped.
        1938-39, 2: 249-256.
1970   Zur Systematick und verbeitung der Gat-
        tung Lampadena Goode and Bean, 1896 (Os-
        teichthyes, Myctophoidei, Myctophidae)
        in Atlantischen Ocean, mit Beschreibung
        einer Neven Art. Ber. Dt. Wiss. Kommn.
        Meeresforsch., 21: 271-289.
1974   Investigations on midwater fish in the
        Atlantic Ocean. Ber. Dt. Wiss. Kommn.
        Meeresforsch., 23: 226-254.
Kulikova, E. B.
1960   Lampanyctids (Genus Lampanyctus) of the
        far eastern seas and the Northwestern
        Pacific (in Russian). Trudy Akad. Nauk
        SSSR, Inst. Okeanol., 31: 166-204.
1961   Material on the lanternfish genus Dia-
        phus (Family Scopelidae) in the western
        part of the Pacific Ocean. Trudy Akad.
        Nauk SSSR, Inst. Okeanol., 43: 5-39.
Legand, M.
1967   Cycles biologiques des poissons mesopel-
        agiques dans l'est de l'Ocean Indien.
        1. Scopelopsis multipunctatus Brauer,
        Gonostoma sp. and Notolychnus valdiviae
        Brauer. Cah. Oceanogr., 5(4): 47-71.
Legand, M., and J. Rivaton
1967   Cycles biologiques des poissons mesopel-
        agiques dans l'est de l'Ocean Indien.
        2. Distribution moyenne des principal
        especes dell'ichtyofaune. Cah. Ocean-
        ogr., 5(4): 73-98.
Lloyd, J. J.
1963   Tectonic history of the south-central
        American Orogan: The backbone of the
        Americans-Tectonic history from pole to
        pole, a symposium. Mem. Am. Ass. Petrol.
        Geol., 2: 88-100.
Lönnberg, E.
1905   The fishes of the Swedish South Polar
        Expedition. Wiss. Ergebn. Schwed. Sub-
        polarexped., 5(6): 1-69.
Lütken, C. E.
1892   Spolia Atlantica. Scopelini museu zoo-
        logici Hauniensis. Bidrag til Kindstab
        om det aabne Haus lakesild eller Scope-
        liner. K. Danske Vidensk. Selsk. Skr.,
        Ser. 6, 7: 221-297.
Mackintosh, N. A.
1937   The seasonal circulation of the ant-
        arctic macroplankton. Discovery Rep.,
        16: 365-412.
1946   The antarctic convergence and the dis-
        tribution of surface temperatures in
        Antarctic waters. Discovery Rep., 22:
        177-210.
Margolis, S. V., and J. P. Kennett
1970   Antarctic glaciation during the Tertiary
        recorded in subantarctic deep-sea cores.
        Science, 170: 1085-1087.
Marr, J. W. S.
1962   The natural history and geography of the
        Antarctic Krill (Euphausia Superba Dana).
        Discovery Rep., 32: 33-464.
Mauchline, J., and L. R. Fischer
1969   The biology of euphausiids. Vol. 7,
        Advances in marine biology, F. S. Rus-

sell and M. Yonge (Eds.), 454 pp. Academic, New York.

Mayr, E.
1963   Animal species and evolution. 797 pp. Harvard University Press, Cambridge, Mass.

McCulloch, A. R.
1915   Biological results of fishing experiments carried on by the F.I.S. Endeavour 1909-1914. 3(3): 104.
1923   Fishes from Australia and Lord Howe Island. 2. Rec. Aust. Mus., 14: 110-142.
1929   A checklist of the fishes recorded from Australia. Mem. Aust. Mus., 5(1): 1-144.

McGinnis, R. F.
1974   Counterclockwise circulation in the Pacific Subantarctic sector of the Southern Ocean. Science, 186: 736-738.
1977   Evolution within pelagic ecosystems: Aspects of the distribution and evolution of the Family Myctophidae. In G. A. Llano (Ed.), Adaptations within Antarctic ecosystems, Proceedings of the Third SCAR Symposium on Antarctic Biology: 547-556. Smithsonian Institution, Gulf Publishing Co., Houston, Tex.

McGowan, J. A.
1971   Oceanic biogeography of the Pacific, Micropaleontology of Oceans, B. M. Funnell and W. R. Riedel (Eds.), 3-73. Cambridge University Press, New York.

McIntyre, A.
1967   Coccoliths as paleoclimatic indicators of Pleistocene glaciation. Science, 158: 1314-1317.

Mead, G. W.
1966   Family Chlorophthalmidae. Fishes of the Western North Atlantic. Mem. Sears Fdn. Mar. Res., 1(5): 162-189.

Mead, G. W., and F. H. C. Taylor
1953   A collection of oceanic fishes from off Northeastern Japan. J. Fish. Res. Bd. Can., 10(8): 560-582.

Midttun, L., and J. Natvig
1957   Pacific Antarctic Waters. Scient. Results Brategg Exped. 1947-1948, 3: 1-130.

Mileikovsky, S. A.
1971   Types of larval development in marine bottom invertebrates, their distribution and ecological significance: A re-evaluation. Mar. Biol., 10(3): 193-213.

Moreau, E.
1888   Le scopele de Verany, Scopelus veranyi. Bull. Soc. Philomath., Paris, Ser. 7, 12(3): 108-111.

Moser, H. G., and E. H. Ahlstrom
1970   Development of lanternfishes (Family Myctophidae) in the California current. 1. Species with narrow-eyed larvae. Bull. Los Angeles County Mus. Nat. Hist., Sci., 7: 1-145.
1972   Development of the lanternfish Scopelopsis multipunctatus Brauer 1906, with a discussion of its proposed mechanism for the evolution of photophore patterns in lanternfishes. Fish. Bull., 70(3): 541-564.
1974   Role of larval stages in systematic investigations of marine teleosts: The

Myctophidae, A Case Study, Fish. Bull., 72(2): 391-413.

Muromtsev, A. M.
1963   The principal hydrological features of the Pacific Ocean. Translated from Russian, catalogue 753: 1-417. Israel Program for Scientific Translations, Jerusalem.

Nafpaktitis, B. G.
1968   Lanternfishes of the genera Lobianchia and Diaphus in the North Atlantic. Dana Rep., 73: 1-131.

Nafpaktitis, B. G., and M. Nafpaktitis
1969   Lanternfishes (Family Myctophidae) collected during cruises 3 and 6 of the R/V Anton Bruun in the Indian Ocean. Bull. Los Angeles County Mus. Nat. Hist., Sci., 5: 1-79.

Nafpaktitis, B. G., and J. R. Paxton
1968   Review of the lanternfish genus Lampadena with a description of a new species. Los Angeles County Mus. Contr. Sci., 138: 1-29.

Newell, B. S.
1966   Seasonal changes in the hydrological and biological environments off Port Hacking, Sydney. Aut. J. Mar. Freshwater Res., 17: 77-91.

Norman, J. R.
1930   Oceanic fishes and flatfishes collected in 1925-1927. Discovery Rep., 2: 261-370.
1937   Fishes. Rep. BANZ Antarct. Res. Exped., Ser. B, 1(2): 50-88.
1938   Coast Fishes. 3. The antarctic zone. Discovery Rep., 43: 1-105.

O'Day, W. T., and B. G. Nafpaktitis
1965   A study of the effects of expatriation on the gonads of the two myctophid fishes in the North Atlantic Ocean. Bull. Mus. Comp. Zool. Harv., 136: 77-89.

Orren, M. J.
1966   Hydrology of the southeast Indian Ocean. Invest. Rep., 55: 1-35. S. Afr. Dept. of Commerc. and Ind., Div. of Fish., Durban.

Ostapoff, F.
1965   Antarctic oceanography, Ecology and biogeography of the antarctic, In J. V. Mieglem, P. van Oyes, and J. Schell (Eds.), Monogr. Biol., 15: 976-126. A. Junk, The Hague.

Pappenheim, P.
1912   Die Fische der Deutschen Sudpolar Expedition 1901-1903. 2. Die Fische der Antarctic aug Subantarctic. Dt. Sudpol. Exped., 13: 162-182.
1914   Die Fishen der Sudpolar Expedition 1901-1903. 2. Die tiefse Fishe. Dt. Sudpol. Exped., 15(2): 160-200.

Parin, N. V.
1970   Ichthyofauna of the epipelagic zone. Translated from Russian, catalogue 5538, 205 pp. Israel Program for Scientific Translations, Jerusalem.

Parr, A. F.
1928   Deepsea fishes of the order Iniomi from the waters around the Bahama and Bermuda Islands, with annotated keys to the Sudidae, Myctophidae, Scopelarchidae, Evermanellidae, Omosudidae, Cetomimidae, and

Rondeletiidae of the world. Bull. Bing-
ham Oceanogr. Coll., 3(3): 1-193.

Paxton, J. R.
1967a A distributional analysis for the lanternfishes (Family Myctophidae) of the
San Pedro Basin, California. Copeia,
2: 422-440.
1967b Biological notes on southern California
lanternfishes (Family Myctophidae).
Calif. Fish Game, 53(3): 214-217.
1972 Osteology and relationships of the lanternfishes (Family Myctophidae). Bull.
Los Angeles County Nat. Hist. Mus., Sci.,
13: 1-81.

Pytcowitz, R. M.
1968 Water masses and their properties at
160°W in the Southern Ocean. J. Oceanogr. Soc. Jap., 24(1): 8-15.

Rafinesque, C. S.
1810 Indice d'Ittiologia Siciliana. Messina,
70 pp.

Regan, C. T.
1913 The Antarctic fishes of the Scottish
National Antarctic expedition. Trans.
R. Soc. Edinb., 49: 229-292.
1914 Fishes, Natural History. Br. Antarct.
Terra Nova Exped. 1910, Zool. 1(1):
1-54.

Reid, J. L.
1965 Intermediate waters of the Pacific Ocean.
Johns Hopkins Oceanogr. Stud., 2: 1-85.

Reinhardt, J. C. H.
1837 Ichthyologiske bidrag til den gronlandska
fauna. Dansk Vid. Selsk. Afh. Copenhagen, 6: 107-111.

Richardson, J.
1844- Ichthyology of the voyage of H.M.S.
1848   Erebus and Terror, Zoology of the voyage
of H.M.S. Erebus and Terror, J. Richardson and J. E. Gray (Eds.), 2: 1-139.

Rochford, D. J.
1960 The intermediate waters of the Tasman
and Coral seas. 1. The 26.80 sigma t
surface. Aust. J. Mar. Freshwater Res.,
11(2): 127-147.

Roper, C. F. E.
1969 Systematics and zoogeography of the
worldwide bathypelagic squid Bathyteuthis (Cephalopoda, Oegopsida). Bull.
U.S. Natn. Mus., 291: 1-210.

Rothschi, H., and L. Lemasson
1967 Oceanography of the Coral and Tasman
seas. In H. Barnes (Ed.), Oceanography
and Marine Biology, Annual Review, 5:
49-97. George Allen and Unwin, London.

Savage, J. M.
1960 Evolution of a Peninsular Herpetofauna.
Syst. Zool. 9: 184-212.

Savage, J. M., and M. C. Caldwell
1965 Studies on Antarctic oceanology. Biological stations occupied by the USNS Eltanin: Data summary, cruises 1-13, Report:
87 pp. University of Southern California, Los Angeles.
1966 Studies in Antarctic oceanology. Biological stations occupied by the USNS Eltanin: Data summary, cruises 14-16, Report:
31 pp. University of Southern California, Los Angeles.
1967 Studies in Antarctic oceanology. Biological stations occupied by the USNS Eltan-

in: Data summary, cruises 22-24 and
26-27, Report: 36 pp. University of
Southern California, Los Angeles.

Schell, I. I.
1968 On the relation between the winds off
Southwest Africa and the Benguela Current and Agulhas Current penetrations in
the South Atlantic. Dt. Hydrogr. Z.,
21: 109-117.

Scripps Institution of Oceanography, Woods Hole
Oceanographic Institution
1969 Physical and chemical data from the Scor-
pio expedition in the South Pacific
Ocean, USNS Eltanin cruises 28 and 29.
SIO Ref. 69-15: 89 pp. La Jolla, Calif.

Shackleton, N. J., and J. P. Kennett
1975 Paleotemperature history of the Cenozoic
and the initiation of Antarctic glaciation: Oxygen analyses in DSDP sites
277, 279, and 281. In J. P. Kennett et
al. (Eds.), Initial reports of the Deep
Sea Drilling Project, 29: 743-755. U.S.
Government Printing Office, Washington,
D. C.

Shannon, L. V.
1966 Hydrology of the south and west coasts
of Africa. Invest. Rep. 58: 1-24. S.
Afr. Dep. of Commer. and Ind., Div. of
Sea Fish., Durban.

Shih, Ç.
1969 The systematics and biology of the family Phronimidae. Dana Rep., 74: 1-99.

Smith, J. L. B.
1933 New Myctophid fish from South Africa.
Trans. R. Soc. S. Afr., 21: 125.

Sverdrup, H. U., M. W. Johnson, and R. H. Fleming
1942 The oceans. 1087 pp. Prentice-Hall.
Englewood Cliffs, N. J.

Taft, B. A.
1963 Distribution of salinity and dissolved
oxygen on surfaces of uniform potential
specific-volume in the South Atlantic,
South Pacific, and Indian oceans. J.
Mar. Res., 21: 129-146.

Tåning, V.
1928 Synopsis of scopelids in the North Atlantic. Vidensk. Meddr. Dansk Naturh. For-
en., 86: 49-69.
1932 Notes on scopelids from the Dana expedition. 1. Vidensk. Meddr. Dansk Naturh.
Foren., 94: 125-446.

Tebble, N.
1960 The distribution of pelagic polychaetes
in the South Atlantic Ocean. Discovery
Rep., 30: 161-300.

Trunov, I. A.
1968 New observations on the distribution of
Electrona rissoi (Cocco) and Diaphus
ostenfeldi Tåning (Myctophidae). Vopr.
Ichtiol, 4: 595-598.

Viglieri, A.
1966 Oceans: Limits, definitions, and dimen-
sions, Encyclopaedia of Oceanography,
1021 pp. R. W. Fairbridge (Ed.), Rhein-
hold, New York.

Voronina, N. M.
1968 The distribution of zooplankton in the
Southern Ocean and its dependence on the
circulation of water. Sarsia, 34: 277-
284.

Waite, E. R.

1904   Addition to the fish fauna of Lord Howe
        Island, 4. Rec. Aust. Mus., 5(3): 135-
        186.
1911   Fishes. 2. Rec. Canterbury Mus., 1(3)
        157-272.
1916   Fishes. Australas. Antarct. Exped.,
        Ser. C., 3(1): 59-62.
Walsh, J. J.
1969   Vertical distribution of antarctic phyto-
        plankton. 2. A comparison of phyto-
        plankton standing crops in the Southern
        Ocean with that of the Florida Straits.
        Limmol. Oceanogr., 14(1): 86-94.
Warren, B. A.
1970   General circulation of the South Pacific,
        Scientific Exploration of the South Pa-
        cific, W. S. Wooster (Ed.), 33-49, Na-
        tional Academy of Sciences, Washington,
        D. C.
Watkins, N. D., and J. P. Kennett
1971   Antarctic bottom water: Major change in
        velocity during the late Cenozoic be-
        tween Australia and Antarctica. Science,
        173: 813-817.
Whitley, G. P.
1931   Studies in Ichthyology. 4. Rec. Aust.
        Mus., 18(3): 96-133.
1932   Studies in Ichthyology. 6. Rec. Aust.
        Mus., 18: 321-348.
1941   A lanternfish from Macquarie Island.
        Aust. Zool., 10(1): 124
1943   Ichthyological notes and illustrations.
        2. Aust. Zool., 10: 167-187.
1953   Studies in Ichthyology. 16. Rec. Aust.
        Mus., 23(3): 133-138.
Wisner, R. L.
1963   A new genus and species of myctophid
        fish from the south-central Pacific
        Ocean, with notes on related genera and
        the designation of a new tribe, Electron-
        ini. Copeia, 1: 24-28.

1971   Descriptions of eight new species of
        myctophid fishes from the eastern Pacific
        Ocean. Copeia, 1: 39-53.
Wooster, W. S.
1970   Eastern boundary currents in South Pacif-
        ic, Scientific Exploration of the South
        Pacific, W. S. Wooster (Ed.), 60-68.
        National Academy of Sciences, Washington,
        D. C.
Wooster, W. S., and P. Gilmartin
1961   The Peru-Chile undercurrent, J. Mar.
        Res., 19(3): 212-220.
Wooster, W. S., and J. L. Reid
1963   Eastern boundary currents, In M. N. Hill
        (Ed), Ideas and observations on progress
        in the study of the seas, The sea, 2:
        253-280. John Wiley, New York.
Wüst, G., and A. Defant
1936   Schichtung and Zikulation des Atlant-
        ischen Ozeans. Dt. Atlant. Exped. Meteor
        1925-1927, Rep. 6, Atlas. 103 pls. Ber-
        lin.
Wyrtki, K.
1960a   The antarctic circumpolar current and
        the antarctic polar front. Dt. Hydrogr.
        Z., 13(4): 153-174.
1960b   Surface circulation in the Coral and
        Tasman sea. Tech. Pap. 8: 1-44. Div.
        of Commonwealth Sci. and Ind. Res. Organ,
        Melbourne, Australia.
1963   The horizontal and vertical field of
        motion in the Peru Current. Bull.
        Scripps Inst. Oceanogr., 8(4): 313-346.
1967   Circulation and water masses in the east-
        ern equatorial Pacific Ocean. Int. J.
        Oceanogr. Limnol., 2: 117-147.
Zugmayer, E.
1911   Diagnoses des poissons nouveaux provenant
        des campganes du yacht Princess Alice
        1901-1910. Bull. Inst. Oceanogr. Monaco,
        193: 1-14.